Contents

M
A
H

Ele

S.W
B.Sc.

GRANADA
London Toronto Sydney New York

Granada Publishing Limited — Technical Books Division
Frogmore, St Albans, Herts AL2 2NF
and
36 Golden Square, London W1R 4AH
866 United Nations Plaza, New York, NY 10017, USA
117 York Street, Sydney, NSW 2000, Australia
100 Skyway Avenue, Rexdale, Ontario, Canada M9W 3A6
61 Beach Road, Auckland, New Zealand

Copyright © 1982 S.W.R. Cox

ISBN 0 246 11717 6

First published in Great Britain 1982 by Granada Publishing

British Library Cataloguing in Publication Data

Cox, S.W.R.
    Microelectronics in agriculture and horticulture.
    1. Microelectronics 2. Farm mechanization
    I. Title
    621.381'71   TK874
Printed in Great Britain by Mackays of Chatham Ltd.

Granada ®
Granada Publishing ®

TO MONICA

Foreword

By John E. Moffitt, C.B.E., F.R.Ag.S

Director, Hunday Friesians, Hunday Electronics, Hunday Museum,
West Cumberland Farmers' Co-operative
Managing Director of Cattle Breeders Services Ltd
Managing Director of Cattle Breeders Services (Embryos) Ltd
Vice President, British Friesian Society
Member of Governing Body, British Society for Research in
Agricultural Engineering

As a farmer born and bred from three generations of farmers on both sides of my family and certainly no mathematician, one might ask how on earth I became interested in electronics.

Electronics has a strange fascination for most of us; it is rather like the supernatural to the lay person because it can apparently do the impossible. How many of us, if we have a problem, immediately say, 'Ah well, electronics will do it for us', without really thinking very hard about it!

My initiation to the subject started exactly that way, with the problem of how to feed accurately and identify 300 cows without the need to refer to names and charts. Electronics solved this problem for me then and it can accomplish very much more today.

There is no need to have an extensive knowledge of the subject although it does help. What we must do, however, is be imaginative about its applications and make sure we know how to get the best out of the equipment and programs.

What Sidney Cox has done in this excellent book, is to help us to understand more fully the technology, and project the likely developments and use we can make in the future with electronics. As Deputy Director of NIAE and previously Head of the Control and Instrumentation Division of the Institute, he has the ideal background and practical experience in agriculture to inform us of the potential the subject can offer the industry.

I am convinced that whether we like it or not, electronics and computers will rule our lives completely in five years, and we have to be no more afraid of it than our forefathers were at the birth of the internal combustion engine. Soon every farmer who runs an efficient business will need one or more computers and all dairy farmers will be monitoring food intake and performance. Future tractors, combines, sprayers and almost every implement will be equipped with electronics to help us do our job more economically and efficiently.

Sidney Cox has covered electronics in agriculture and horticulture very effectively indeed and this book will, I am sure, be a must for the bookshelves of all progressive farmers and growers.

Preface

The emergence and development of microelectronics devices over the past decade have already changed the way in which we do many things at our work and in our social and domestic lives. It has radically extended our capabilities and horizons and the process continues at an ever-increasing rate. Month by month we read of newer, smaller and yet more powerful devices produced in 'Silicon Valley' and elsewhere. Journals regularly forecast the changes to be wrought by the impending 'information revolution', which will follow from the progressive extension and improvement of high-speed data transfer between widely distributed computers. We will have more — and more immediate — access to data banks, weather forecasts, programme packages enabling us to undertake complex technical and economic analyses from our local terminal, and so on. We will move towards a more dispersed work force in many areas of employment, relying on data links rather than energy-intensive travel.

All of these things seem likely, if not inevitable, and the agricultural and horticultural communities will be affected probably as much as any sector of the population. Indeed, the capability of the new electronics and communications technology to cater for dispersed sites is particularly advantageous to the primary food producers and the service groups who support them through installation, maintenance, contract and advisory work in many ways.

Agriculture and horticulture are often thought to be conservative by those outside these industries. Certainly, the complex biological and economic spheres within which the farmer and grower seek to operate profitably are risky enough to deter impulsive experiment. Nonetheless, there has been a substantial increase in the amount and variety of electronic equipment on farms and horticultural enterprises in the last few years, fostered by the farmers' and growers' needs to acquire more precise and timely information on their inputs and outputs, in order to survive in an increasingly competitive world. The number of meetings, journal articles and discussions on farm electronics leaves no doubt about the wide interest in this subject among farmers, growers, advisers, staff and advanced students in agricultural and horticultural colleges, manufacturers and suppliers of agricultural and horticultural machinery and the manufacturers of industrial instrumentation, computing and control equipment who wish to gauge the agricultural and horticultural market for their existing or future products.

This book has been written as an introduction first to electronic instrumentation and then to microelectronics itself — what it is and what it does, to its existing uses in agriculture and horticulture, and to likely developments in the technology and its applications over the next decade or two. Because microelectronics is still evolving rapidly the text adheres as closely as possible to fundamental concepts, which are relatively unchanging. It is intended to provide the necessary grounding for the specialists in other fields — agriculture and horticulture, mechanical engineering design, management, marketing, etc. — who realise that they must understand and be responsive to the opportunities presented by microelectronics technology. Manufacturers of industrial electronics equipment will not need the introduction to instrumentation and microelectronics but it is hoped that they will be stimulated by the considerable existing and potential market for instrumentation, computing and control in agriculture and horticulture, while taking note of some of the special hazards presented by the environment in these two industries.

Perhaps the greatest hazard to the supplier of farm electronics at present is the customer's fear that the equipment will let him down at a crucial time. The questions most frequently heard from farmers and growers are about the reliability of the equipment and what remedies will be available if or when it breaks down. Designers and suppliers of mechanical equipment have had to face these questions over many years, of course, and the agricultural engineering industry has done much to mitigate the effects of breakdowns through dealer networks, design standardisation and customer training in first-line maintenance. Maintenance courses at educational and training centres are well-established, too. There can be little doubt that the rate of further uptake of microelectronics equipment in agriculture and horticulture will be highly dependent on its overall reliability, the provision of first-line diagnostic facilities, the ready availability of replacement parts or modules, the accessibility of maintenance and calibration services and the existence of suitable training courses for farm staff. Although matters of detailed design and servicing are outside the scope of this book, as are details of training courses, general comments on vulnerability, standardisation, calibration and maintenance of equipment are included at appropriate points.

Expertise in mathematics is not required of the reader. Formulae are quoted in a few places, as a convenient supplement to the text for those who are sufficiently familiar with the symbolism. International Standard (SI) Units have been employed throughout. These are listed, with others, in Appendix 1.

The author wishes to acknowledge his indebtedness to the many colleagues at the NIAE, other research stations, the Ministry of Agriculture's Agricultural Development and Advisory Service (ADAS), and in engineering industry, both in the U.K. and overseas, who have supplied material for and made helpful comments on the text. Special thanks are due to Doreen Croxford for her care in producing and checking the typescript.

S.W.R. Cox

1 The Elements of Industrial Electronics

1.1 Introduction

The origins of electronics technology can be traced back to physics research in the late nineteenth century. Working with metal plates and filaments sealed in evacuated glass enclosures, J.J. Thomson and others showed that negatively-charged electrical 'particles' (electrons) could be emitted from the surface of the metals under the influence of light (photoemission) and heat (thermionic emission). The electrons so liberated could be collected by a second plate (anode), made electrically positive with respect to the first (cathode) by connecting both to a battery. The two-electrode assembly therefore generated a one-way electrical current. This was the basis of Fleming's thermionic 'diode' (1904). Then in 1907 de Forest introduced the thermionic triode, which had a third, 'grid' electrode between the anode and cathode. Small changes of grid potential relative to the cathode controlled the current through the triode, which thereby provided a means to amplify weak radio signals and, more generally, to drive electrical control systems from low-power electrical inputs.

The thermionic vacuum tube developed and became the mainstay of radio and TV communications, audio systems and an increasing range of industrial monitoring and control applications. Operating in an 'on-off' switching mode it also became the heart of the first digital computers in the late 1940s and early 1950s. Nevertheless, it had some major limitations, especially for mobile applications. In particular, it required substantial electrical power to heat the cathode and equally substantial voltages to provide the necessary anode-cathode potential.

Fortunately, research into the electrical properties of crystals, also originating in the nineteenth century, led to the development of the low-power, low-voltage semiconductor devices that have replaced thermionic components in most areas of electronics.

Semiconductor materials are intermediate between metal conductors, with their abundant supply of free electrons and insulators (dielectrics) which have no conduction electrons. Semiconductors have a limited supply of charge carriers of both signs (positive and negative), which move in opposite directions under the influence of applied voltages.

Semiconductor diodes had been used in radio circuits for years before Bardeen, Brattain and Shockley invented the transistor — the semiconductor equivalent of the triode — in 1947. However, the transistor's three electrodes, termed 'emitter', 'collector' and 'base' (the base corresponding to the triode's grid), provided the means to design semiconductor oscillators, amplifiers, switches, etc. and in doing so began a giant step forward for industrial electronics.

The effects of these developments on agriculture and horticulture will almost certainly be far-reaching because — as pointed out in the preface — farmers and growers need more precise and timely information in order to maintain the profitability of their enterprises in an increasingly competitive world. Electronics equipment helps them towards this end in several ways. First, it provides them with tools for measuring what is going on in their operations, for co-ordinating measurements of different kinds and for processing the resultant data into the forms that are needed for soundly based operational and management decisions. Second, it makes possible automatic control of many processes, through the combination of electronic instruments with electrical and electronic controllers, thereby releasing men from laborious and repetitive tasks and often improving on manual control of these tasks. Third, it offers the power and flexibility of the digital computer for management purposes, including the collection and analysis of data from many sources, both on-farm and external.

The greater part of this book is concerned with the first two aspects of electronics, namely automatic monitoring and control of agricultural and horticultural operations. Later chapters deal with specific sectors of agriculture and horticulture, discussing the quantities to be measured or controlled and the equipment available or under development for these purposes. However, there are many features common to electronic measurement and control systems which can be more usefully treated collectively at the outset. The present chapter is concerned with these features generally, while chapter 2 introduces microelectronics and its influence on them. A glossary of electronics is included at the end of the book.

1.2 Electrical and electronic measurements

Electronic measurement and control systems require an electrical input so, clearly, unless the quantity to be measured or controlled is itself electrical (e.g. current or voltage) the first step is to employ a device which responds to that quantity by generating an electrical output dependent in some way on the magnitude of the quantity. Such devices are generally called *sensors* or *transducers*. Sometimes they are designed to respond in an 'on-off' mode to changes in the magnitude of the quantity, as a thermostat does in relation to temperature changes, or as some bar-code readers do at sales checkout points in relation to reflected light. This form of action is common in industrial process control, where electrical and electronic switches combine with electronic 'logic' (see 1.3.8, 1.4.1 and 1.5) to determine automatically which way and at what rate a process will proceed. Of more immediate concern here are the devices which respond in an 'analogue' mode, viz. the magnitude of the physical or chemical quantity is converted to an electrical output of proportionate amplitude. These devices provide inputs to a wide range of measurement and control systems.

Details of some commonly used analogue transducers are given in 1.3 but first it is important to consider some general aspects of their performance and that of complete instruments, together with equally important considerations of errors and accuracy. These topics are fundamental to much of what follows in later chapters. General discussion of the performance of transducers and instruments must centre on the design and environmental factors which limit their working range and the fidelity of their analogue output. The most important of these factors are outlined in the first of the two following sub-sections. The second sub-section deals with errors that can arise when successive stages of a measurement system are coupled together, including the sometimes significant effect of transducers on the quantity to be measured. It also includes consideration of the averaging and sampling errors which result from changes in the measured quantity from point to point or moment to moment. These are nearly always important factors in agricultural and horticultural measurements. It concludes with reference to the vital subject of calibrations and that often misunderstood term, 'accuracy'.

1.2.1 *Transducer and instrument performance*

Some characteristics of transducers and instruments generally need little explanation and will only be briefly discussed here.

Range   All of this equipment has an upper and lower limit of working response to the measured variable even though the lower limit is sometimes effectively zero (as in many weighing devices, for example).

Linearity   Ideally, if a change x occurs in the measured quantity at any point within the working range of a transducer or complete instrument the response of the equipment − y, say − should be the same throughout the range, i.e. its sensitivity y/x = constant.

If this ratio is not constant at the transducer stage it may be required to correct it within the system. Some methods of linearisation will be outlined later in this chapter.

Manufacturers usually quote the claimed linearity of their transducers and instruments in terms of the maximum non-linearity at any part of their range, as a percentage of their full range output.

Hysteresis   Significant hysteresis in a transducer or instrument produces an output which at any instant is partly dependent on what has gone before. The amount and sense (i.e. increase or decrease) of prior change in the measured quantity determines the size of the effect. This has an obvious influence on the repeatability with which a given magnitude of the quantity can be measured, as well as on the linearity of the output. Hysteresis is usually measured at half-range and quoted as a percentage of full-range output.

Drift   This can be systematic or random. For example, manufacturers' leaflets often quote the systematic effect of changes in ambient temperature on the output of their equipment in terms of per cent change per deg C. 'Creep' may occur in load sensing devices under prolonged loading. Changes in ambient conditions generally and other influences, such as fluctuations in mains voltage supplies, may produce unpredictable drifts overall. In addition, ageing of the equipment will result in further, longer-term drifts, some of a systematic character and some not. Repeated manual or automatic drift correction is therefore a necessary feature of many instrumentation systems.

Precision The precision (or, in standard metrology usage, discrimination) with which measurement can be taken will depend on the form in which the output of the measuring instrument is presented and the steadiness of that output. In general, analogue meters (i.e. those with moving pointers or other index marks) offer lower precision than digital meters which display three or more decimal digits. However, precision must never be confused with accuracy, which is discussed in sub-section 1.2.2.

The dynamic characteristics of measuring equipment deserve more attention here because they have an important influence on the fidelity of the equipment's response when the measured quantity varies at more than a critical rate, and this fact is sometimes overlooked.

Rigorous analysis of instrument dynamics requires full mathematical treatment but several basic concepts can be introduced in fairly simple terms.

Frequency response The notion of frequency response will be familiar to anyone with an interest in high fidelity audio recording and playback. Manufacturers strive to produce amplifiers and other audio components with a substantially flat (even) response across the audio spectrum. Producers of transducers and output devices generally (but not necessarily) seek the same flat response over a wide frequency range. Many such devices behave dynamically in the same way as a solid mass on a spring, with some degree of viscous damping. Without damping the suspended mass will oscillate at its 'natural' (or resonant) frequency, f_0. Damping has two effects: it progressively diminishes the amplitude of the vibrations and it changes the frequency of the oscillation. Both of these effects increase with increasing damping to the point at which oscillation just fails to occur. At this point the damping is said to be critical, beyond it the system is 'over-damped'.

Figure 1.1 shows a set of frequency response curves which summarise the dynamic performance of a transducer or instrument which acts as a mass/spring/damper system. It depicts the response of the system to fluctuations in the measured variable of the same maximum amplitude but of varying cyclic frequency, from one-hundredth of the natural frequency f_0 to three times that frequency. Given an ideal, flat frequency response the output of the system would be the same at all frequencies and the amplitude ratio (defined as the output at any frequency, divided by that at $f = 0$)

would always be 1.0. It can be seen that the ideal response holds up to about 0.1 f_0: thereafter it rises and falls according to the value of ζ, the damping ratio, which is another ratio — the actual amount of damping present in the system as a proportion of its critical damping. It is clear that a damping ratio of about 0.6 provides the flattest response and in fact a ratio of just under 0.65 maintains a level response within ± 2% up to a frequency ratio of 0.65. For these reasons reference to natural frequencies and damping is often found in manufacturers' specifications of their equipment.

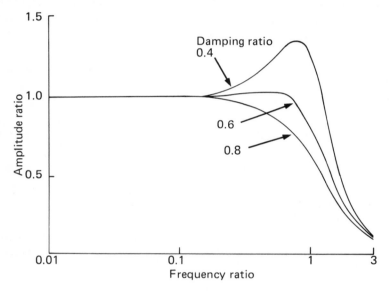

Fig. 1.1 Frequency responses of a 'mass/spring/damper' system at three levels of damping (NIAE).

Of course, measured variables rarely vary cyclically in an obligingly sinusoidal (pure tone) manner; normally they fluctuate in complex ways. However, mathematically it is possible to represent many continuously varying quantities as an infinite series of harmonically related sinusoidal disturbances. Fortunately, it is not necessary for practical instruments to respond to more than about twenty of the lower harmonics: this is usually adequate to reproduce a given waveform with an accuracy of a few per cent. Nevertheless, the highest harmonic must lie in the level section of the instrument's frequency response for even this degree of fidelity, therefore it must be remembered that for acceptable accuracy of measurement of a quantity varying in a complex manner the fundamental frequency of the variation may have to be less than $0.65 f_0/20$.

Another point to note is that there is a phase lag in the equipment's response at each frequency and this increases with frequency. Conveniently, the resultant phase error is minimal near the $\zeta = 0.65 = f/f_0$ conditions for optimal frequency response.

Transient responses   Sometimes it is necessary to obtain as close an estimate as possible of a sudden change in a quantity. This change may be an impulse of short duration, or a sudden step change. Fig. 1.2(a) shows the time response of the mass/spring/damper system to a transient at three different levels of damping, while fig. 1.2(b) depicts the response to a step change of amplitude 1.0 at four damping ratios.

Once again, a damping ratio of 0.65 gives an acceptable response to an impulse. The step response of a system with some degree of damping will initially overshoot the final value by about 7% at the worst: generally, the figure will be lower, since step changes are not absolutely instantaneous in the real world.

Time constant   Not all transducers and instruments respond as mass/spring/damper systems, some are non-resonant and respond to step changes exponentially.

The mathematical expression of their response to a sudden positive or negative change, X, in the measured quantity can be written

$$x_t = X \left[1 - \exp\left(-t/\lambda\right)\right]$$

Where x_t is the change in output of the equipment t seconds after the change in the measured quantity, e is the Naperian constant, 2.718 and λ is known as the time constant or lag coefficient of the equipment. The term $\exp\left(-t/\lambda\right)$ diminishes with time, so that x_t steadily approaches X. It reaches 0.63X when $t = \lambda$ seconds and 0.998X (i.e. effectively its final output) when $t = 6\lambda$.

The response of a thermometer transferred from a colder to a hotter position, or vice versa, follows this pattern.

1.2.2 *Errors, accuracy and calibration*

The preceding sub-section has outlined some common sources of error in the output of transducers and other measuring equipment even when these are used within their specified range of operation,

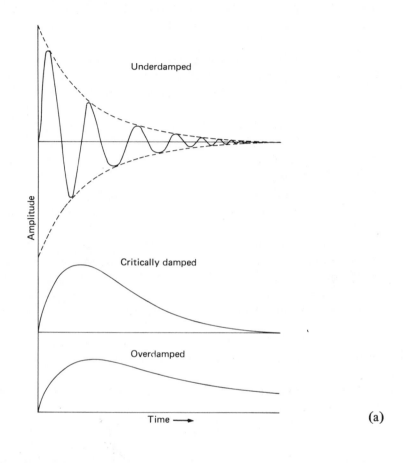

(a)

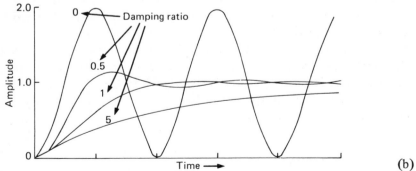

(b)

Fig. 1.2 (a) Transient responses of a 'mass/spring/damper' system at three levels of damping. (b) Step responses of the system at four levels of damping (NIAE).

i.e. excluding non-linear response due to some form of overload. In this connection careful matching of separate stages of an instrumentation system is sometimes necessary to avoid overdamping or overload in the system. In electronic measuring instruments, too, electrical noise derived from sources of electrical and electromagnetic power can be superimposed on the measured signal, with resultant distortion of the signal, or even overload of the system. These two points are discussed briefly here.

Impedance matching Manufacturers' specifications for transducers, amplifiers and other components of measurement systems frequently quote the electrical input or output impedance of their equipment, as do suppliers of audio equipment. The objective is the same, i.e. to enable the user to minimise the effect of interconnections on the amplitude and frequency responses of the system. Electrical impedance embraces both electrical resistance, which is constant whether the circuit is handling d.c. or a.c. signals, and electrical reactance, which varies with the frequency of the signal. Impedance is frequently quoted in units of resistance (ohms). This may mean that the resistive component is dominant but it may also imply an average figure (taking account of resistance) over the working frequency range. Referring again to audio electronics, this is common practice in the case of loudspeaker specifications.

Electrical noise In electronics terminology, noise is an unwanted signal of a random or systematic nature. It can be picked up as radiated interference from local electrical and magnetic fields generated by electrical equipment, or as mains-borne interference, entering via the power cable. It can also be generated in the measuring equipment itself in various ways, both electrical and electronic.

Minimisation of electrical noise and its effects is partly a matter of instrument design and partly of proper selection, siting and installation of equipment.

Another class of error, mentioned in the preamble to this section, is brought about by the inevitable interaction of the transducer itself with the conditions of measurement. For example, an electrical resistance bulb placed in room, container or elsewhere to measure the temperature of the surrounding medium will disturb the preexisting heat balance in two ways. First, unless it was originally at the same temperature as the medium it will supply heat to or extract heat from the latter by an amount depending on its thermal capacity (defined as its mass multiplied by its specific heat) and the original temperature difference. Second, as a resistive element with a current passing through it, it absorbs power from its electrical supply by an amount I^2R_t, where I is the current and R_t the resistance of the element at a particular temperature. This creates an additional and continuing supply of heat to the medium while the element is energised. Apart from these two forms of disturbance the electrical leads to the element will be responsible for transfer of heat energy by thermal conduction either into or out of the medium.

The effects just outlined are usually small in relation to the heat capacity of the surrounding medium, fortunately, but possible interactions of this kind must always be borne in mind when a measurement system is being set up. Care is needed to ensure that flowmeters do not impede substantially the liquid or gaseous flow that they are intended to measure, that the movement of diaphragms in pressure sensors does not materially affect the pressure in a medium, and so on. Some transducers are superior in these respects to others.

Next, there are sampling and averaging errors to consider. Two aspects of this subject will be taken in turn.

Instrument averaging errors   Many of the electrical and electronic instruments used to measure rapidly fluctuating quantities display a single reading which is intended to provide a representative value for the average level of the measured quantity. Commonly this reading is derived by averaging the square of the electrical signal over a given time and taking the square root of this average. Given a sinusoidal signal the outcome is the true r.m.s. (root mean square) value encountered widely in a.c. voltage, current and power measurements at mains frequencies. This is not so when the waveform is cyclic but non-sinusoidal. If it is not cyclic but of a random nature statistical theory defines the duration of the averaging period that is necessary to obtain a particular degree of confidence in the reading. Qualitatively, the sample must contain a sufficiently representative sample of those frequency components in the signal which are of interest. The required averaging time is therefore determined by the lowest frequency in this range.

Sampling errors   In materials as inhomogeneous as soils, crops and animal products it is unlikely that a representative magnitude of a particular variable will be obtained from a measurement at one place. The same may apply to measurement of the aerial environment in buildings. These circumstances call for multiple sampling and the application of the simple statistical routines to quantify errors that are commonly incorporated in pocket calculators for technical and scientific use. The routines are based on the concept of a 'normal' distribution of errors and lead to the determination of the standard deviation, σ, of a set of measurements, defined as follows:

Given a set of n measurements of a quantity, $x_1, x_2, \ldots \ldots, x_n$, the mean value of these readings, $\overline{X} = (x_1 + x_2 + \ldots + x_n)/n$. Then

$$\sigma^2 = \frac{(x_1 - \overline{X})^2 + (x_2 - \overline{X})^2 + \ldots + (x_n - \overline{X})^2}{n - 1}$$

Broadly, the probability that an individual measurement will lie outside the range $\overline{X} \pm 2\sigma$ is about 1 in 18 and the probability reduces to about 1 in 500 outside $\overline{X} \pm 3.1\sigma$. The value of σ therefore characterises the dispersion or scatter of the measurements. Sometimes this scatter about the mean value is expressed in terms of the coefficient of variation, expressed as a percentage and defined as $(100\,\sigma\,/\,\overline{X})\%$. A third useful measure is the standard error of the mean value, defined as $\alpha = \sigma/\sqrt{n}$.

Means and standard deviations can also be derived from a time series of measurements at a particular point or on a particular object; for example, the weight of an animal at different times.

Another aspect of sampling statistics in measurement relates to counting of radioactive particles or photons of electromagnetic energy, such as X-rays, which are generated randomly. Statistical analysis shows that for the mean count rate to lie within $\pm 2\%$ of the 'true' mean, with 95% probability, the sampling time must be long enough to allow the accumulation of 10 000 counts.

Accuracy and calibration   Users of industrial measuring equipment usually want to know what accuracy it is capable of; manufacturers in their turn quote the accuracy claimed for their equipment. It is pertinent to consider what this means.

The accuracy which can be attributed to a measurement is the numerical difference between the measured value of a quantity and the 'actual' value of this quantity, insofar as it can be determined. Therefore, the only way to establish the accuracy which can be attained by an instrument is to use it to measure a quantity whose magnitude is known to a much higher degree of accuracy: in other words it must be calibrated (verified) against some standard.

Calibration can be direct, through reference to a reproducible condition of matter. For example, thermometers can be checked against the melting point of pure ice (0°C) or the boiling point of pure water at 100 kPa atmospheric pressure (100°C). More commonly, calibration is indirect, either via comparison with the indication of a 'standard' instrument or via measurement of a standard test-piece or pieces, such as a calibrated set of weights. Ultimately these indirect standards must themselves be referred to national or international standards of measurement, directly or at one or more further removes. The objective is to establish 'traceability' to ultimate standards, wherever possible, and services such as the British Calibrations Service exist to promote this link with instruments in the hands of the industrial user.

The output drifts mentioned in 1.2.1 make it unlikely that the calibration of any instrument will remain substantially unchanged throughout its working life, therefore regular rechecks are essential to ascertain that it is within the claimed accuracy or, if it is not, that it can be readjusted to restore its calibration. It is obviously preferable to check at more than one point in the instrument's range, since this reduces reliance on its continuing linearity, or whatever other law relates its output to its input.

The foregoing outlines the basis for the quotable accuracy of an instrument but it does not guarantee the accuracy of a particular 'working' measurement. Away from the precise conditions of a calibration or check reading care is needed to ensure that the instrument is working within its specified dynamic and environmental limits. Too rapid variations in its readings may introduce observer error too, as everyone who has attempted to read the weight of a restive animal with a spring balance will vouch.

Beyond this lies the statistical scatter which has been referred to earlier. There is no virtue in measuring a quantity to 0.1% accuracy if in practice conditions are such that the coefficient of variation of a set of such measurements is, say, 2%. In agriculture and horti-culture, as in many other industries, the need for accuracies better than ± 1% is rare and ± 2% or 3% is often adequate. Unjustifiable precision in presentation of measurement results is all too common. The high precision of pocket calculator and computer displays has done much to spread this practice, unfortunately.

1.3 Some common types of transducer

After the generalities of the preceding section some consideration is now given to several types of transducer which are found extensively in industrial applications of electronics. More specialised sensors and transducers are dealt with at appropriate points in later chapters.

1.3.1 *Displacement, load and pressure*

Agriculture and horticulture provide many applications for transducers which measure the distance or change of distance between two points. Examples are the height of a point on a machine above ground level; the level of a liquid relative to the top or bottom of its container; the distance moved by a component under pressure or as the result of thermal expansion. Perhaps the most common application is in weighing. In traditional mechanical weighers, where the load is supported by a spring or other flexing element, the movement of the mechanism under load is sufficient to drive a mechanical pointer across the calibrated scale. The spring balance has provided this mechanism in animal weigh crates and it is a relatively simple matter to measure the extension of the spring by mounting a linear displacement transducer in parallel with it, to provide an analogue of the weight (cf. 6.4.1).

Another class of transducer with a variety of applications employs electrical strain gauges to measure the deformation of a diaphragm, cantilever beam or other deformable component under load. The movements involved in this case are very small, in the micrometre (μm) range. Load and pressure, torque and acceleration are quantities commonly measured with transducers of this type. In agriculture and horticulture weighing again provides the most frequent application for them.

A description of the many varieties of linear and strain gauge transducers would require more space than can be allowed here but the salient features of the most common types can be given.

Linear displacement transducers The simplest and cheapest form of these employs a linear resistance element (wirewound or conductive plastic) in a cylindrical housing, within which a coaxial rod slides on low-friction bearings, moving a sliding contact over the element. The relative movement between the housing and the rod under displacement is thereby converted to a change in resistance which can be measured electrically. Ranges of from 10 mm to 500 mm are available. It is also possible to convert the linear displacement into proportional rotary displacement by a suitable mechanical linkage (for example, a pulley and cord) and to use a rotary potentiometer as the sensing element, i.e. converting linear to angular displacement. The resistance sensor can be remarkably reliable under farm conditions although its sliding contact is obviously at risk under adverse environmental conditions and rapid displacements, and electrical contact noise can be generated. Its resistance is usually a few kilohms and the resistance change per unit displacement of the core rod is uniform, but in the wirewound types the movement of the sliding contact over the resistance winding produces small step changes in the output.

The resistance transducer usually carries a d.c. current and so produces a low voltage d.c. analogue of the linear or angular displacement. Contactless linear transducers, energised by a.c. currents, are also widely available. These are generally more expensive than the resistance transducers but they are less susceptible to the environment and have infinite resolution (i.e. no step changes in output with deflection). The most common type is the L.V.D.T. (linear variable differential transformer) which contains three coaxial coils, side by side, surrounding a small coaxial rod or armature of magnetisable material which is normally located symmetrically with respect to them. The centre coil is a.c. energised and the two outer ones connected differentially. Any movement of the core rod along its axis from the central, null position unbalances the system and an out-of-balance voltage signal appears between the extremities of the two outer coils. The phase of the out-of-balance signal relative to that in the energised centre coil depends on the direction of displacement of the core from its null position. This phase-sensitive output is then used to measure the magnitude and direction of the core displacement.

Many of these transducers are fed from low voltage d.c. supplies and contain an oscillator which energises the centre coil at a frequency in the kHz range. The out-of-balance voltage is rectified by a phase-sensitive 'demodulator' and converted to a d.c. analogue voltage. L.V.D.T.s are usually made with ranges between ± 2.5 mm (minimum) and ± 25 mm but some long-stroke types have a one-way range of over 500 mm. The core rods, running on low friction bearings, can be free-moving or spring-loaded. Like the resistive transducers their low voltage d.c. output is compatible with semiconductor circuits. The manufacturer's specification of the performance of a d.c./d.c. unit might look like this:

Input	6V d.c., 20 mA
Range	± 10mm
Linearity	0.3% of F.R.O. (full range output)
F.R.O.	± 2V d.c. Maximum a.c. ripple 1% rms
Zero drift	0.5 μm/^{0}C
Span temperature drift	−0.01%/^{0}C
Temperature range	−20^{0}C to +100^{0}C
Frequency response	0 to 200 Hz (−3 dB)

i.e. the response has diminished by 3 decibels or 30% at 200 Hz (see appendix 1, Units and Standards, for more information on the decibel scale). Note, though, that the mass of the armature rod is not negligible in all circumstances. If attached to a lightweight system it could affect the system's dynamics.

A particularly valuable application for the L.V.D.T. is level measurement, with the armature rod attached to a float. Because the coupling between the rod and the coil assembly is magnetic and not by physical contact it is possible to protect both against attack by the fluid by suitable encasement and, conversely, to prevent them from contaminating the liquid. The latter point is important if an L.V.D.T. is used to measure the level of milk in a recording jar, for example.

The non-contacting capacitance displacement sensor is sometimes employed for linear displacements up to about 300 mm. This is also a.c. energised and may have an oscillator/demodulator circuit built in, to give a low-voltage d.c. output. Like the L.V.D.T. it has a more than adequate frequency response for most agricultural and horticultural applications.

Strain gauged force transducers   Strain gauges are available in many sizes, shapes and materials, with different electrical resistance, strain sensitivity, maximum extensibility, working temperature range and fatigue life. Some are of metal alloy, in wire, foil or vacuum-deposited form; others are of silicon semiconductor material, with high-strain sensitivity but high-temperature coefficients of both sensitivity and resistance, too. Also, the strain sensitivity of semiconductor gauges is less linear in relation to strain level than it is in wire and foil gauges. Semiconductor gauges can be invaluable where transducers of very small size and high frequency response are required but the metal alloy gauges are predominant in industrial transducers. However, the vacuum-deposited gauge which first appeared in the late 1970s found ready application in agriculture for weighing poultry (cf. 6.4.2). The gauge is deposited directly onto a cantilever beam (plate 1), unlike other alloy types, which have to be attached to the deformable component of the transducer.

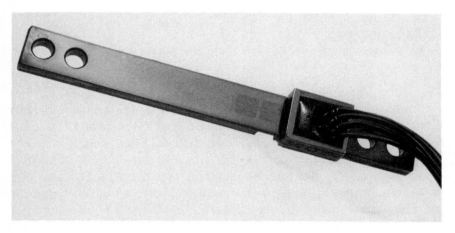

Plate 1 Cantilever beam with vacuum-deposited strain gauge assembly, for load measurement. Full range, 50N (5 kg weight, approximately) (Salter Measurement Devices Ltd.).

In general, proper choice and the location of these gauges is a job for the specialist, and so is their bonding to the surface of the deformable component and the protection of both gauges and leads with protective coatings. The performance and reliability of the transducer under industrial conditions depends critically on the care taken in their design and assembly.

Force or load cells which respond to tensile or compressive loads exist in nearly as many different forms as strain gauges themselves. Their common features are multiple gauges, connected in a Wheatstone bridge configuration, which is energised by a low voltage a.c. or d.c. supply. Some degree of compensation against side loads and the effect of temperature change is usual, although this does not absolve the user from limiting both sources of error as far as possible.

Specifications vary widely too, but a load cell for 2 tonne F.R.O. could be expected to offer the following performance (percentages in terms of F.R.O.).

Output at 12V excitation   20 mV
Hysteresis   0.2%

Creep (20 min)	0.05%
'Zero' temperature drift	0.05%/$^{\circ}$C
Span temperature drift	0.05%/$^{\circ}$C
Temperature range	-20°C to $+80^{\circ}$C
Maximum overload	50%
Deflection	0.05 mm maximum
f_0	5 kHz
Fatigue life	10 M cycles minimum
Input and output resistance	$240\Omega \pm 5\Omega$

Pressure transducers A diaphragm with full bridge configuration (four active gauges) is the basis of most of these transducers. This can be in the form of an integrated circuit, with a silicon diaphragm of very small dimensions (even as little as 1 mm diameter) but except in applications where this size and the accompanying high value of f_0 are needed alloy-gauged transducers still have an advantage in performance.

The pressure or differential pressure that can be measured with the latter types ranges from a few kPa to hundreds of MPa — sufficient for most needs. They can be used in any attitude. Their performance specification follows that of the strain-gauged force transducer very closely, however, so a separate list of performance features is not called for here.

Diaphragms can be made part of a variable capacitance too, and some very sensitive low-pressure transducers are of the capacitance type. They employ the same a.c. oscillator/demodulator system as the capacitance displacement gauges already mentioned and give a.c. output of several volts, full range. They are also very tolerant of overload (even up to $\times 1000$ on low ranges).

1.3.2 *Speed and acceleration*

This sub-section deals with the motion of machines and components. Air speed and fluid flow are dealt with in 1.3.3.

Accelerometers Although physically the logical progression is from displacement, through velocity to acceleration, accelerometers will be discussed first since strain gauges once again feature in many of the standard types and essentially accelerometers are force measuring devices, comprising a mass, m, which experiences a force F = am when subjected to acceleration a. In the strain-gauged accelerometer the mass is mounted on a flexible element with gauges bonded to the element or attached to the mass at one end and to the accelerometer casing at the other. Both types cover a wide range of accelerations.

Piezoelectric accelerometers operate in a slightly different way. Acceleration of the mass creates a pressure on a block of crystalline piezoelectric material, which responds by generating an analogue voltage between opposite faces, in the direction of the pressure. These accelerometers have high values of f_0 and are very robust but they generate very little current and must be coupled to amplifiers with high input impedance to avoid being overloaded, particularly at low frequencies.

Two important characteristics of the accelerometer are its sensitivity to accelerations along other axes than the 'sensitive' axis and its total mass. The transverse sensitivity of the accelerometer is usually expressed as a percentage of its response in the direction of the sensitive axis and this should not amount to more than about 3%. The mass of the unit is important because this determines the minimum mass of the system to which it can be attached without disturbing the conditions that it is intended to measure.

Speed measurement   Electronic integration of the output signal from an accelerometer will provide an analogue of the velocity of the moving object to which the accelerometer is attached. In theory this is an elegant method but in practice distortion and drifts can set in.

Vehicle speed is in any case conveniently determined from the angular revolutions of a wheel of known radius moving over the ground with no slip and no skid. A tachometer generator coupled to the wheel will give an analogue output proportional to speed with good linearity. Alternatively, a digital signal proportional to wheel revolutions can be obtained by deriving a pulse from an electrical or electronic switch of some kind each time the wheel rotates through a given angle. There are several ways of doing this electronically. First, and probably the most reliable in an agricultural context, a magnetic inductance transducer can be used. This is in the form of an iron-cored winding, energised at low voltage, which senses the proximity of metal objects and will respond to the individual teeth on a gear wheel rotating with, or coupled to, the main wheel. Equally, if there is not too much dust and dirt about, a disc with peripheral holes or slots can be used to interrupt the light transmitted from a small semiconductor light source to a semiconductor photodetector, such as a silicon diode. These are only two of the ways to generate pulses at a rate proportional to the rate of revolution of the wheel. A ratemeter circuit will convert the pulse train into a voltage analogue of forward speed. The pulses themselves allow direct accumulation of distance travelled too, of course.

Unfortunately, in agricultural field work a wheel may sometimes skid and slip, while on soft and bumpy ground its effective radius is not always evident and motion is sometimes more upward or downward than forward. The Doppler radar speed meter has been taken up in agriculture to overcome these limitations in operations where control of forward speed is important, as in crop spraying. However, this is a specialised measurement, and is dealt with in the context of field crops (3.3.1).

1.3.3 *Flow of liquids, gases and solids*

Liquids   Measurement and control of liquid flow is often crucial in the spraying operation mentioned in 1.3.2 and in several other areas of agriculture and horticulture. The ways of sensing liquid flow are legion, although the choice narrows considerably when sensors without an electrical output are excluded. In this general section two aspects of the topic are dealt with. Both concern flow in pipes.

Several widely used methods for measuring volumetric flow in pipes can employ the same electrical sensor, namely the pressure transducer, set up to measure differential pressure. The pressure difference to be measured, Δp, is generated by a constriction placed in the pipe — the orifice plate and the venturi nozzle, or one of its derivatives, being the most familiar. The pressure is measured just upstream and downstream of the constriction and its relation to the volumetric flow, Q, is of the form

$$Q = k\sqrt{\Delta p}$$

where the 'constant', k, depends on the cross-sectional areas of the pipe and the constriction, and on the density of the fluid. The absence of moving parts in the tube is an advantage of this type of metering equipment: it is also capable of working reliably with a wide range of fluids (liquids or gases). All the constrictions introduce some head loss but the more complex (and expensive) types introduce lower loss than the simple orifice plate. The value of k depends in part on the 'coefficient of discharge' of the constricting element. This can usually be obtained with moderate accuracy (say, ± 2%)

from standard specifications.

The propeller or turbine metering element provides another insertion method for determining volume flow. In essence, a section of pipe is fitted with a freely rotating propeller which fills the cross-section of the pipe as far as practicable and the speed of rotation of this element is measured. This measurement is usually made via an external magnetic pick-up, which registers the passage of each propeller blade. If the blades are of non-magnetisable material the tips are fitted with a small magnet, to achieve the same effect. The resulting low-voltage pulse train is immediately compatible with semiconductor electronics, which generate the flow-rate signal. This type of meter is compact and creates little head loss. It is not very accurate at low flow rates, because bearing friction begins to dominate at the bottom of its range. From around 20% of F.R.O. upwards it becomes fairly linear; the actual range and linearity depend very much on size and detailed design. The relationship between volume flow and the rate of rotation of the propeller depends on the viscosity of the liquid, which may change substantially with temperature. The calibration will change with usage, too, which makes it imperative to recalibrate the instrument from time to time, for maximum accuracy of measurement.

Air and other gases   Pipe flowmeters of the orifice and venturi type can be used for gases as well as for liquids, as already mentioned. Turbine meters can also be employed but in order to produce the right amount of impulsion to the blades the central section of the pipe is blocked by a boss-shaped housing which contains the propeller bearing. The gas is constrained to flow through the remaining annular space, with the result that it is accelerated before it reaches the blade tips. These metering elements provide better accuracy than orifice plates and are far less dependent on recalibration. In particular, they are not much affected by the viscosity of the gas.

Rotating vane or cup anemometers with electrical pick-ups are widely used to determine the mean air speed in buildings and in the open over durations of about 3 s. They cover a wide range of flows. If velocity (speed and direction) is of interest, as it can be in the glasshouse sector, for example, then the Dines anemometer can be used. Mounted on a swivelling support, and attached to a wind vane, its total-pressure sensing tube faces into the wind, in the manner of a pitot-static tube. Wind speed can then be determined from the difference between this pressure and the static pressure sensed by circumferential holes in the vertical housing of the swivelling support. This differential pressure can be determined by a sensitive electrical capacitance gauge, forming part of a micromanometer. The wind speed is derived from an expression of the form $v = k\sqrt{\Delta p}$, i.e. similar to the relationship already given for volumetric flow of liquids. Even when this gauge is located remotely from the pitot-static tube and coupled to it by flexible pipes the whole system is responsive enough to measure wind gusts of short duration, say, about 3 s. Wind direction can be measured with the aid of a rotary resistance potentiometer, capable of 360^0 of rotation, coupled to the swivelling assembly.

Solids   Measurement of the bulk flow of solids along pipes or conveyor lines reintroduces the subject of weighing. Conveyor-belt weighers are common in industry and have application in agriculture (see 7.3.1). Electronic load cells, supporting sections of conveyor, provide analogue signals which can be used to measure mass flow rate and, after integration, to measure total mass transported. Similarly, batch or continuous weighing devices can be inserted in pipeline distribution systems to provide instantaneous and total mass flow.

1.3.4 *Temperature*

Apart from weight and volume, temperature is the most widely measured quantity in agriculture and horticulture. In the present context, methods of measuring temperature introduce three types of transducer, two contacting types and one not.

The thermocouple   The principle of this traditional and still used method of measuring temperature is widely known, although some of the precautions needed to achieve the expected accuracy are not. Given a circuit made of two dissimilar metals, A and B, connected $B - A - B$, with a sufficiently sensitive meter connected to the two free ends of metal B, the meter will record a voltage if the temperature at the junction $B - A$ is different from that at $A - B$. Symbolically, if the two temperatures are T and T_0 (the latter being treated as a stable reference temperature), the output voltage can be represented by

$$V = \alpha (T - T_0) + \beta (T - T_0)^2$$

where α and β are constants, depending on the metals A and B.

It will be seen from the presence of the squared term that the relationship between voltage and temperature difference is not linear. Further, the output is usually low: if $T - T_0 = 100°C$, V is likely to be only 4 to 5 mV. High-gain amplifiers may be needed to make the output compatible with the requirements of the measurement system and this introduces the possibility of noise pick-up. Accurate control of the reference junction temperature no longer necessitates use of the traditional vacuum flask (filled with pure granulated ice in equally pure water, regularly stirred, to provide a steady $0°C$) or other laboratory apparatus. Electrical means of generating a reference temperature have been developed. The thermocouple is therefore not to be ignored where small size and high speed of response, high accuracy or high temperature capability are the criteria. Thermocouple wires of pure metal or alloy with closely controlled properties are available for both low- and high-temperature work. Their reliability depends much on the care with which the fused junction between the two metals has been made, though, and great care is also needed to ensure that the measuring equipment does not bring into the circuit unknown, thermally generated voltages arising from other contacts between dissimilar metals. If the use of long leads is inevitable one of the methods devised for reducing thermally or electrically induced error must be incorporated. In all, equipment and cables designed for thermocouple work should be used for reliable results.

Resistance elements   The positive temperature coefficient of the electrical resistivity of pure metals and many alloys provides the basis of temperature sensors with a resistance that increases significantly with temperature. The resistance, R, of the element at a temperature, T, relative to its value, R_0, at temperature T_0 is given approximately by

$$R/R_0 = 1 + a (T - T_0) + b (T - T_0)^2$$

where a and b are constants for the particular metal. Again, a squared term is present but over small temperature ranges (i.e. $T - T_0$ is small) the linear relationship $R/R_0 = 1 + a (T - T_0)$ can be employed. R_0 and T_0 usually refer to $0°C$. The materials most widely used for this form of resistance thermometer are platinum and nickel. The approximate value of a for the two materials is +0.004 and 0.007, respectively. The former provides transducers with more repeatable characteristics; for example, elements can be interchanged with less than ± 0.1°C effect on readings. However, its resistance change/°C is smaller. Typically, $R_0 = 100\Omega$ and $R_{100} = 140\Omega$.

These transducers are mounted in a protective metal sheath, usually cylindrical, with a sealed exit for the wires. The time constant depends on the thermal mass of the whole assembly. Electrically, the elements are normally connected in a Wheatstone bridge circuit, with compensation for long leads, if required (see fig. 1.3). The output is readily matched to microelectronics systems.

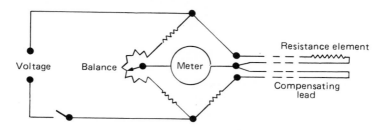

Fig. 1.3 Resistance thermometer bridge, with compensation for long leads (NIAE).

Thermistors This semiconductor element, like the thermocouple, can be very small, with a high speed of response, and it consumes less power than the metal wire element, therefore it is less likely to cause errors through self-heating. On the other hand, although interchangeability can be within ± 0.2°C, the highest accuracy must not be expected from thermistor measurements. Ignoring self-heating, its resistance-temperature relationship is approximately

$$R/R_0 = \exp\left[b\left(1/T - 1/T_0\right)\right]$$

Note: T is in degrees kelvin, where K = °C + 273
The temperature coefficient of resistance of the element is large and it decreases with T, non-linearly. For example:

$$T = 273 \text{ K} \quad (0°C), R = 3\ 000\Omega$$
$$T = 323 \text{ K} \quad (50°C), R = \quad 700\Omega$$
$$T = 373 \text{ K} \quad (100°C), R = \quad 200\Omega$$

Thermistor circuits are commonly but not always of the bridge type. The length of leads is less important than it is for metal wire elements because of the higher resistance of the thermistor beads.

Remote sensors At times measurement of temperature through contact with the material concerned is not convenient nor is it always desirable. Remote sensing of the temperature of liquid and solid surfaces is possible with infra-red radiation thermometers, which have become less expensive, following the development of new sensors. However, their optical systems are much more expensive than resistance thermometer or thermocouple elements, which retain a clear economic advantage when multipoint measurements are required. More details of the infra-red thermometer are given in 1.3.7.

1.3.5 *Humidity*

The humidity of the environment is a pervasive influence in every industry and most operations in agriculture and horticulture are affected by it. Unfortunately, it is not easy to measure continuously and even those sensors which can be used for this purpose are liable to deteriorate in use, unless regularly maintained and checked. Some are irreversibly damaged if subjected to a polluted atmosphere.

The most reliable sensor is still the wet and dry bulb unit. Contemporary forms employ two platinum resistance elements, one of which is fitted with a thin, woven sleeve which dips into a supply of distilled water in the usual way. The necessary minimum air flow over this wet bulb (3 m/s) is achieved with a small d.c. motor. The use of platinum resistance temperature sensors ensures accurate measurement of the often small temperature depression of the wet bulb relative to the dry bulb. This is essential if the relative humidity of the atmosphere is to be calculated with any degree of accuracy over the range of values required by industry. Even so, for meaningful measurements maintenance of a clean wick and water supply are essential. This demands attention at weekly intervals, if not more frequently.

Many attempts have been made to produce a sensor more convenient for industrial measurement and a variety of commercial elements are available. None has yet shown the ability to survive unchanged in the harsher industrial environments, which include most agricultural environments. Apart from the treated hair element which changes length with humidity and is often the basis for electrical humidistats (of long time constant), several forms of electrical absorption element exist. These contain thin hygroscopic films which exchange moisture with the surrounding atmosphere and so exhibit changes in their electrical resistance or capacitance. Since the hygroscopic film stores very little moisture their effect on the environment is minimal.

Resistive types are fairly slow in response, with a time constant of about 30 s, unless they are aspirated, in which case this reduces to about 5 s (the lag is longer when the humidity is decreasing). Their resistance-R.H. relationship is non-linear and hysteresis may amount to ± 3% R.H. Their output must also be temperature-corrected.

Capacitance elements have a time constant of less than 1s and are relatively temperature insensitive. Microelectronic techniques have been applied to the construction of some of these sensors, to produce an hygroscopic polymer film only 1 μm thick. Thin gold film electrodes complete the capacitor structure. The increase of capacitance with water absorption is linear within ± 1% R.H. up to 80% R.H. Other features are:

Hysteresis (from 0 to 80% R.H. and back)	Better than ±1%
Temperature coefficient	About 0.05% R.H./°C
Temperature range	−40°C to +80°C
Claimed accuracy up to 80% R.H. at 20°C	± 2% R.H.
Water uptake at 100% R.H.	About 0.5 μg
Heat uptake	About 0.2J/°C

Above 80% R.H. some drift occurs over a period of hours. This can be overcome by incorporation of a tiny heater/temperature sensor, which is automatically controlled by an electronic circuit to heat the sensor to the temperature at which it only reads 75% of the actual R.H. at any R.H. The sensor is therefore never required to measure an R.H. higher than 75% and the performance is essentially as given above. The penalty is greater cost and the generation of a moderate amount of additional heat in the surrounding medium (10 mW in still air). The need to maintain the required excess temperature over ambient also imposes a limit on the air speed over the sensor (10 m/s). The size of the sensor described above is shown in fig. 1.4. Normally it is protected by a sintered filter, which can be seen at the end of the probe in plate 2. This keeps dust from the sensor and reduces the condensation which can settle on the unheated probes under some conditions. The presence of the filter increases the time constant of the system, of course. Other types of probe are available, including one for measuring the E.R.H. (equilibrium relative humidity) of the air in bulk powders and granular materials. A calibrator box with two saturated salt solutions is available for the unheated sensor, too. This provides check readings at about 12% R.H. and 97% R.H., the actual values depending on the ambient temperature. Electrically, the sensor is a low-voltage device. Its associated electronic circuit is mounted in the probe and this can be connected to a temperature/ R.H. indicator (plate 2). Temperature is measured with a thermistor in the probe. Multiple probes can be connected in turn to one indicator, via a switching box.

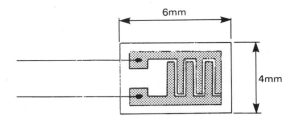

Fig. 1.4 Capacitative, thin film humidity sensor (Vaisala Ltd.).

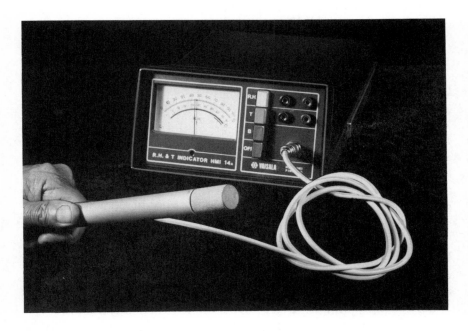

Plate 2 Humidity probe incorporating the sensor in fig. 1.4, with meter (Vaisala Ltd.).

Relative humidity can also be measured very reliably to an accuracy of about 1% R.H., by bringing the air into contact with a polished metal surface which is cooled until the moisture in the air is deposited as dew on the surface. The temperature of the surface is measured at the dew point, together with that of the ambient air. From these two values the R.H. of the air can be calculated. One meter working on this principle employs a Peltier effect (thermoelectric) element to cool the mirror cyclically. The onset of dew is detected by a silicon photocell which monitors reflected light from an LED (light emitting diode). The instrument memorises the dewpoint on an initial (fast) cycle and updates the value on subsequent slower cycles. Since this dynamic measurement is based on detection of a sharp change in reflectance it is relatively insensitive to the build-up of dust on the mirror, which makes for reliability in dusty environments. Nevertheless, the sensor may be fitted with a sintered filter, to give added protection. Plate 3 shows six of these sensors, with a six-input multiplexer, a meter with temperature and R.H. scales, and separate digital thermometer. An ambient temperature probe is mounted in each sensor. The instrument also incorporates diagnostic features for fault detection. Readings take at least a few minutes, depending on circumstances.

Plate 3 Dewpoint sensors for humidity measurement, with meter and microprocessor-based multiplexer (Protimeter Ltd.).

The dewpoint cell, like the capacitance probe (plate 2), is robust enough to be placed in bulk materials, to provide a value of the E.R.H. there. Both have uses in crop drying and storage (chapter 5).

1.3.6 *Electrical conductivity and pH*

The electrical conductivity of liquids is often measured industrially in order to monitor the concentration of ionisable constituents in solution. In agriculture and horticulture this measurement is particularly associated with measurement of plant nutrients (4.3). Standard commercial conductivity equipment employs a low-voltage a.c. resistance bridge, with the conductivity cell as one arm of the bridge and the output meter scaled in units of conductance. The cell comprises two plane platinum electrodes in a glass or metal holder and, since the electrical conductivity of liquids is very temperature-dependent, it also contains a resistance thermometer. The measurement must be made under a.c. conditions to avoid polarisation effects in the liquid.

The specific conductance of the liquid can be deduced from the measured conductance by multiplying this value by the electrode's stated 'cell constant', which can be derived from the area of the electrodes and their distance apart but is more reliably determined from measurements with a liquid of known conductivity. Accuracies of around ± 2% are attainable, if the cell is kept clean and the electrodes free of air bubbles during measurement.

A companion measurement is often the degree of acidity or alkalinity of the solution, on the pH scale of 1 to 14, which indicates the hydrogen ion concentration, with the neutral point between acidity (low pH) and alkalinity (high pH) lying near pH 7, depending on temperature. The pH sensor comprises two electrode assemblies (the reference and measuring electrodes), each containing solutions of known ionic activity. Some contain platinum resistor elements, to compensate for wide variations in the temperature of the solution. When the probe with the two electrodes is placed in the solution under test the electrical circuit is completed and the output voltage between the reference and measuring electrodes is a measure of the pH of the solution. The time constant is of the order of seconds. Little current can be drawn from this assembly (which may have a resistance of 1000 MΩ) and the meter must have a high input impedance. However, integrated circuit amplifiers of the required impedance are available, to provide the input stage and to match the sensor to succeeding stages of the equipment. Maintenance of the pH electrode assembly according to the manufacturer's instructions is important, if the attainable accuracy of ± 0.1 pH is to be achieved. This maintenance includes replenishment of a 'salt bridge' solution from time to time. The equipment should be calibrated regularly against commercially available standard solutions, too.

1.3.7 *Radiation*

This sub-section deals with equipment for measurement of radiation in the electromagnetic spectrum from long wavelengths (the infra-red and heat bands), through the visible light range (760 nm down to 400 nm) to the X-ray bands. This is another sphere in which the range of commercially available sensors is wide but, in the context of this book, it is possible to concentrate on three classes here.

(a)

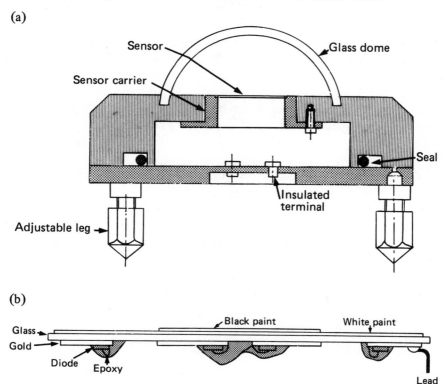

(b)

Fig. 1.5 (a) Solarimeter, (b) diode solarimeter: section of sensor (NIAE).

Thermal detectors These are infra-red detectors which respond thermally to the intensity of the radiation incident on them in an effectively uniform way over a wide band of wavelengths. Solarimeters for measurement of combined sun and sky radiation are an example, with meteorological and horticultural applications. The solarimeter's sensor is based on a blackened disc which absorbs the radiant energy and becomes hotter in the process. Its rise in temperature above ambient provides a measure of radiation intensity. One form of the sensor assembly is a cylindrical metal block on levelling feet (the black disc must lie in the horizontal plane), with a hollow centre and a hemispherical transparent cover of good optical quality in glass or plastic, as shown in fig. 1.5(a). The black disc is painted on the upper, central zone of a larger disc, of thermally insulating material. The remainder of this surface is painted white. Two concentric sets of temperature-sensitive elements are fixed to the underside of the disc, the inner one located within the radius of the black disc and the outer one outside it, beneath the white paint. Both sets are thermally isolated from the case of the instrument by their insulating substrate but the outer array equilibrates with the main body of the instrument through heat exchange by radiation and convection. This design ensures that the difference in temperature between the inner and outer elements is proportional to the radiation intensity with adequate linearity for industrial application.

Apart from linearity, important features of a solarimeter are:

(i) Its cosine response. If the illumination I strikes the black disc at an angle θ to the vertical, the output from the sensor should be proportional to I cos θ. The optical quality of the dome has a strong influence on this feature.

(ii) Its azimuth response, which should be as uniform as possible throughout 360°.

(iii) Its time constant, which affects its response to changes in radiation levels.

(iv) Its overall ambient temperature coefficient, which should be −0.2%/°C or better.

(v) Its sensitivity to radiation intensity, in V/Wm^{-2}.

Figure 1.5(b) shows a section of a sensor which employs silicon chip diodes of small thermal mass as temperature-sensitive elements. The forward voltage drop across these diodes when they carry current is temperature-dependent, providing a change of over −2mV/°C. They are mounted on two concentric arrays of vacuum-deposited gold segments, using conductive epoxy resin. The outer and the inner rings, connected serially, form adjacent arms of a resistance bridge.

Use of silicon diodes provides a high output for a solarimeter, at about 150 μV/Wm^{-2}, giving over 15 mV in sunny conditions. However the time constant of about 15 s is rather longer than that of some instruments with lower output, which employ thermocouples.

Infra-red thermometers These are optical instruments which find many applications for remote reading of surface temperatures. They cover the temperature range from +2000°C down to −100°C, corresponding to radiation maxima from 1.5 μm in the near infra-red out to the far infra-red. Three different types of radiation sensor can be used:

(i) The bolometer, which is a thin, blackened strip of platinum which changes resistance when it absorbs the radiation falling on it. It forms one arm of a Wheatstone bridge circuit, which contains a similar strip, screened from the radiation, in an opposing arm. The latter compensates for changes in ambient temperature. The bolometer has a flat response over an exceptional range of wavelengths — from around the red spectrum (0.7 μm) to the far infra-red (2000 μm). and a moderately short time constant (about 10 ms). This wavelength span cannot be maintained in practice if the surface is viewed through a lens or window in the instrument. However, germanium and other materials provide windows to 20 μm and beyond, and this allows low temperature measurement, to well below 0°C.

(ii) The thermocouple or thermistor element, with a similar time constant to the bolometer and a flat response out to about 20 μm.

(iii) The pyroelectric detector. This is a relatively new type, based on triglycine sulphate and related materials. The TGS transducer also has a reasonably flat response to about 20 μm and a time constant of only 1 ms. However, it does not respond to steady illumination since its behaviour is capacitive rather than resistive. It therefore requires some means of modulating the radiation that falls on it. Commonly a rotating shutter is used.

These instruments are able to scan small or large surface areas and possess a temperature resolution of ± 1°C or better. Unfortunately, their readings can be several degrees in error unless the ambient temperature and the emissivity of the surface are taken into account.

The emissivity is a fractional number which indicates the amount by which the surface falls short of an ideal 'black body' (i.e. a body that absorbs all the radiation that falls on it). The fraction varies with the wavelength of the radiation and has to be determined experimentally. While this can be done for structural materials, the complexity of biological materials makes the determination difficult in agricultural and horticultural applications.

Photocells   In contrast to the wideband detectors just described there are many photodetectors with less uniform response and more limited spectrally which nevertheless find wide application. Most of the photocells which are employed in the visible wavelengths are of this nature. However, when a measurement has to be made at a particular wavelength — or, more practically, over a narrow band of wavelengths — then non-uniform spectral response is less of a problem. Colour measurements on crop materials often come into this category.

The silicon photodiode is a versatile conductive detector, with a short time constant (1 μs or less) and a usable response from the near infra-red (1 μm) down into the X-ray region. It has many applications relevant to agriculture and horticulture. The germanium photodiode and phototransistor are also widely used. These have a useful visible/near infra-red (1.6 μm) response but greater temperature sensitivity. Polycrystalline, conductive sensors such as cadmium sulphate and cadmium selenide have peak sensitivity in the visible light region. However, they are much slower in response than the silicon and germanium 'junction' devices and they require a higher voltage power supply.

It is convenient here to mention another optical device which can be regarded as part transducer, part electronic circuit element. This is the optocoupler or optoisolator, which transmits an analogue or digital signal from one stage of a system to the next without electrical connection. The light output of a silicon light-emitting diode (LED) is modulated by the incoming signal current and picked up by a photodiode in the same package, which reproduces the original electrical signal. The coupler is compatible with other elements of the microelectronics family (chapter 2). It can transmit analogue signals with a flat frequency response to 1 MHz and beyond. Digital pulses can be transmitted at very high rates. This device helps to overcome problems of electrical insulation and of noise pick-up in the circuits. It is also closely related to the elements used to convey information over fibre optics communications links (section 8.4).

1.3.8 *Event markers and coding devices*

This section would be incomplete without further reference to the class of transducers — mentioned at the start of this chapter — which provide information of the 'on-off' variety to an instrumentation and/or control system. The thermostat is one of the commonest examples; it senses temperature but only generates information at two preset levels, signalling that it is time to switch the heating system on or off. Action can be started or stopped at times determined by many other types of electrical and electronic switch, singly or in combination as decreed by the logic designed into a control system. The magnetic and photo-electric switches which are used to produce trains of uniform pulses for measurement of angular rotation (1.3.2) are equally suitable as event markers, signalling the arrival or departure of a particular object at a particular point in the system being controlled. The list includes ultrasonic switches; electrical conductivity switches; a variety of mechanically operated switches, of course, and time switches, many of which are based on a frequency reference, such as a crystal-controlled oscillator or the mains frequency. Their compatibility with microelectronic circuits is the relevant point here. In fact, most of the electronic switches now available are designed to generate the binary logic '0' and '1' levels (around 0V and 5V) which satisfy the requirements of semiconductor circuits. Those which merely activate a set of electrical contacts present no problems in this respect, either.

Finally in this section, it is necessary to mention digital codes. For many years linear and disc-shaped encoders, carrying digital codes, have been available to the mechanical engineering industry for measurement of linear and angular displacement. More recently, bar codes have become commonplace in shops and elsewhere as means to identify particular objects or classes of object. As a rule, the coding strips are 'read' by an optical or magnetic detector but other interrogation systems are possible. This type of measurement — if it can be called that — has already found application in cattle production (chapter 7) and is likely to extend into other sectors of agriculture.

1.4 Signal processing and output devices

Sections 1.2 and 1.3 have dealt with some aspects of those stages of an electronic system which follow the input transducer and there have been references to output meters in several places. This section is devoted to further information on the intermediate and output stages of electronic measurement and control systems, leading into the more specific discussion of microelectronics and computing in chapters 2 to 8.

1.4.1 *Signal conditioning and processing*

The output from a transducer may have to be 'conditioned' to the succeeding stages in several ways. An impedance-matching operation is sometimes needed, as stated in 1.2, and/or it may be necessary to amplify or attenuate the signal. In other cases it is essential to filter the signal to extract or suppress a frequency component or a band of frequencies. The rejection of electrical noise can be an important consideration. Deliberate non-linearity may be introduced, in order to correct non-linearity in the transducer signal or to modify it to the shape required for later stages. As an example of the last point, it may be required to convert a sinusoidal waveform into a rectangular one as a preliminary to counting the number of cycles within a given time period.

Operational amplifiers   The electronics engineer has standard ways of arranging the above operations on the original signal and many others. In this connection the operational amplifier, in one of its many forms, is ubiquitous — certainly as far as analogue circuits are concerned. This device has two inputs and one output. One of the former provides an inverted output, i.e. the output signal diminishes as the input signal increases and vice versa; the other input is non-inverting. The output also responds to the differential voltage across the two input terminals. In this configuration it provides high rejection of what is called common mode noise. This is the interference which is picked up equally by the two inputs. The common mode rejection ratio (CMRR) is one important feature of these versatile devices, which are to be found in pH meters (1.3.6), bridge circuits, active filters, integrators, power supply regulators and elsewhere. However, the general-purpose 'op. amp.' has its limitations and it may be necessary in some applications to employ those types which have superior performance in terms of fast response, high stability, low input offset voltage and other features. These are more costly, as would be expected.

Analogue to digital converters More and more, analogue signals are converted to corresponding digital signals before further processing. This operation is carried out by an analogue-to-digital (A/D) converter, following preliminary signal conditioning and, sometimes, a sample-and-hold stage. A/D converters, which are in part based on operational amplifiers, may operate in one of several ways, but with essentially the same result. The analogue signal is sampled at some instant, enters the input of the converter and emerges in a 100 μs or less as a set of bits (binary '0's and '1's), corresponding to the amplitude of the signal at that instant. The digital signal is normally taken from a parallel set of output lines, each of which can only be electrically at one of the two binary levels, around 0 V and 5 V respectively. A serial binary output is sometimes available, too. Figure 1.6 shows an A/D converter's function diagrammatically. From the output of this device the transformed signal becomes digital data, which may be processed in one or more different ways, including conversion from parallel binary form to serial binary form, computer processing and D/A (digital to analogue) conversion for output to an analogue display or control unit.

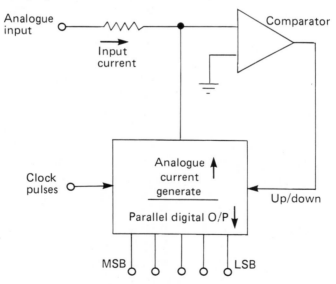

Fig. 1.6 Simplified diagram of an A/D converter with a 6-bit output. In a tracking converter the comparator loop continually follows changes in the analogue input. In a successive approximation converter the input is sampled and held until the conversion is complete, then the next sample is taken. (MSB = most significant bit. LSB = least significant bit) (NIAE).

A/D converters vary in their speed of conversion, their precision (i.e. the number of binary digits, or bits, in their output) and their linearity, which may be as high as ± 0.5%. As with operational amplifiers, this is reflected in their range of costs.

Multiplexers　In many measurement and control systems more than one transducer is involved but it is not necessary to record the outputs of all of them simultaneously. This makes it possible to connect each sequentially to the A/D converter via a multiplexer with multiple input channels (typically 4, 8 or 16), which switches from one to the next electronically at a rate determined by 'clock' pulses. The connection on each channel is made by solid-state switches which are 'addressed' in turn by binary signals generated by the clock, the 'break' from one channel being completed before the 'make' to the next. These switches may have an 'on' resistance of 200Ω or more, so the incoming signals need to be conditioned to the 1V level at least. To avoid interactions between channels both leads from each input are brought into the multiplexer separately (i.e. no common earth or ground connection is involved). The output from the multiplexer goes to a buffer op. amp., to reduce common-mode interference.

Digital circuits introduce another set of common building blocks, too. There is a family of 'logic' elements or 'gates', possessing multiple inputs and single outputs which adopt only the two logic levels '0' and '1'. The AND unit gives an output '1' if all its inputs are at '1', while the NAND (Not AND) unit gives an output '1' unless all its inputs are at '1'.

Similarly, an OR and a NOR produce output '1's if any one or none of their inputs is at '1' respectively. With building blocks like these, complex decision-making circuits can be designed for computing and sequential control systems. In effect they control data flow according to the logic built into the circuit. They also feature in the adding circuits which are basic to computer calculations. Accompanying these elements are bi-stable electronic switches (flip-flops), which change from one state to the other each time they receive an input signal of the required kind. Groups of these can perform counting and memory functions. In consequence they are built into digital counters and timers, and feature extensively in computer circuits. In the latter context they work in conjunction with logic gates and with the crystal-controlled timers which provide the clock pulses needed for control of the sequence of computations and data transfers. By these means, and in association with bulk storage devices (1.4.2), digital processing of the data derived from the original signals can be performed.

1.4.2 *Output devices*

Meters and recorders   The analogue meter with its moving pointer has survived the arrival of the digital display and has enduring merits, including advantage from the ergonomics standpoint. The electrical panel meter of this type can provide a clear indication of the magnitude of a quantity in relation to its full range or some set point. It also has economic advantage in many cases, is reasonably robust, and is ordinarily compatible with semiconductor electronics. Much the same comments apply to chart recorders of several different types. It is interesting to note that the design freedom provided by low-voltage liquid crystal displays has led to the development of displays which exploit the ergonomic advantages of analogue presentation. In the computer field, colour graphics on the VDU (visual display unit) are becoming steadily more economic for general industrial monitoring of processes, too, through developments in hardware and software.

Little needs to be added on the subject of digital displays. The reliability and readability of these displays has greatly improved since their introduction commercially, when they tended to be poor on both counts in the agricultural environment.

Printers have made rapid strides in capability as a result of the development of digital computers – and particularly of the microcomputer. Their speed of operation has increased and is still increasing; at the same time their cost is decreasing. Nevertheless, they remain a vulnerable item in adverse environmental conditions if they are not well protected and maintained. This also applies to keyboard input/output instruments, such as the teletype.

Data storage   Among the peripherals of computer systems, bulk data stores have become increasingly important, since the microcomputer has developed such powerful capabilities. The cheapest (but slowest) is still magnetic tape, on which binary data are entered and played back in serial form. Faster response is possible from magnetic discs, with data storage on concentric tracks, which can be searched radially and circumferentially. These are now complemented by magnetic bubble stores, in which the data, represented by minute regions of magnetised material (the 'bubbles'), can be stacked and unstacked (serially) without the need for motor-driven read/ write heads. The bubble memory is also non-volatile, i.e. it does not lose the data if there is a power failure.

All of these devices — and new forms under development — are capable of storing megabytes of information in a unit of modest size, each byte (8-bit group) representing a number or an alphabet character.

Control devices   If the output from an electronic monitoring system is required to control an operation some form of power amplification is nearly always needed. The control of a.c. electric motors for conveyors, fans, etc., is a case in point. Most of these operate at mains voltages, which are not compatible with microelectronics devices. Fortunately, the thyristor or silicon controlled rectifier (SCR) bridges this gap. This is a semiconductor triode (3-terminal) device, with two of the terminals termed the 'anode' and the 'cathode' respectively, borrowing the labels from the thermionic gas-filled triode which preceded them. In the thyristor the 'gate' supersedes the former 'control grid'. The device is switched on by a low voltage, low power signal applied between the gate and the cathode and once it is conducting the gate loses any further control until the anode current has been reduced to zero. This happens naturally when the thyristor is in series with the a.c. mains power supply and the electrical load (motor, etc.). It will switch off at the end of a half-cycle of the mains current and will only switch on again another half-cycle later if the gate voltage is still present. Motors, solenoids, etc. will not operate as intended on alternative half-cycles of power, of course, therefore the triac has been developed. This is, in effect, two thyristors in parallel, which conduct in opposite directions, thereby allowing on-off switching of both half-cycles. If the gate signal fluctuates at mains frequency and its phase is changed relative to that in the load circuit the triac can be made to conduct for a fraction of each half cycle, the amount depending on the phase change. This method is used to regulate the power in heavy electrical loads but it produces sharp changes in the load current which can create noise problems in local electronics circuits unless power filters are employed.

For smaller loads, such as low-voltage solenoid actuators or electrical relays, simple power transistors may be employed as switches, driven from their base (control) terminals between the fully off and fully conducting states. This avoids high power dissipation in the transistors themselves.

Heat dissipation is an important design consideration for all semiconductor power devices. Overheating may cause malfunctions or permanent damage to them.

Thyristors can be used to control power to d.c. loads, too, by 'chopping' a d.c. supply at varying on-off ratios, while the triac is employed in power frequency changers for variable-speed a.c. motor drives.

1.5 Control systems

Several references have been made to control systems in earlier sections. A few words on the terminology employed in this book are included here. The sequential control systems already mentioned in relation to logic are to be found in domestic washing machines, where a step in the process may await a '1' signal from a temperature sensor, or in a conveyor system for cattle (7.3.1).

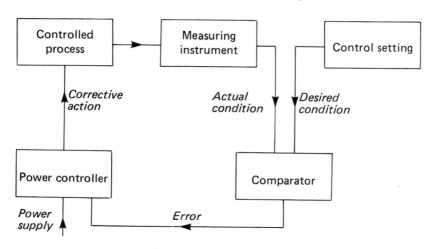

Fig. 1.7 The elements of a closed-loop control system (NIAE).

Open-loop control systems drive the output elements (motor, heater, etc.) from the setting of a manual control of some kind without reference back, and rely on both the calibration of the electronics system and the influence of the environment to be substantially constant. On the other hand, closed loop control systems contain a feedback path from the output to the input, as shown in fig. 1.7. The value of the quantity to be controlled is determined with a suitable transducer, whose output is compared with the manual setting. The resultant error, if any, is amplified and fed into the control element, which is thereby adjusted to reduce the error. Suffice to note here that in the real world, where things change unpredictably, it is never possible to reduce the error to zero. However, by the application of control theory it is often possible to reduce errors to acceptable levels under normal operating conditions.

1.6 Maintenance and reliability

The last section in this chapter must be dedicated to the general aspects of reliability of electronic equipment. In particular the farmer or grower with electronic monitoring and control systems must be able to feel confident that the indications of meters and recorders can be relied upon and that the control system will not break down at crucial times. Some things can be done by the designer and manufacturer, who should ensure first that the system is built to withstand the environment in which it will be working and second that it has diagnostic aids which will help the user to determine the existence — and maybe even the location — of a fault. In the latter connection, voltage sensing LEDs can be used with effect to monitor selected parts of the system, for example. Also, instructions on installation, operating procedure, calibration and maintenance need to be clearly and imaginatively set out, using graphics aids, if mishandling of the equipment by the customer is to be minimised. Electronics designs are particularly prone to small modifications arising from deliberate or enforced changes in component designs and specifications. The supplier who does not keep his documentation in good order is storing up trouble for himself and the user.

The customer can practise self-help by choosing suppliers with good servicing and calibration facilities and with the ability to provide a prompt response when needed. Installation should be by people trained in setting up instrumentation systems, too.

The reliability of electronic monitoring and control systems in buildings often depends on the care with which cable runs and terminations are put in. Badly located wiring from sensors can be the reason for pick-up of electrical noise. Instrumentation cables should be kept clear of mains power lines, in general, particularly when the former are working with low currents or voltages. Metal-screened cables may be needed sometimes to reduce electrical interference. It may also repay the extra cost if signal conditioning is placed close to the transducers, thereafter transferring the data to the central monitoring equipment in a relatively noise-immune condition. This may be achieved by analogue amplification and/or impedance matching (low impedance circuits are less liable to suffer inter- ference). A/D conversion in the signal conditioning equipment, followed by serial binary data transmission, also improves noise immunity.

The location and screening of wiring on field machinery is some- times critical in respect of interference as well as vulnerability to mechanical damage. The dangers of damage to electronic equipment through subjection to repeated vibration and shock must not be forgotten, although microelectronics equipment is far less vulnerable than earlier forms of electronics were.

Whether it is in buildings or on farm machinery, the equipment must be well protected against soil, dust, moisture, corrosive fluids, etc. The quality of electrical connectors is important. These can be a common source of unreliability. Some of the transducers themselves are also potential sources of trouble through their exposure to the environment, including mechanical hazards. Malfunction of strain gauges through ingress of moisture has caused trouble on many occa- sions – even with transducers which had survived in other industries. The agricultural and horticultural environment is exceptionally severe!

The advice offered to would-be purchasers of electronics equip- ment by specialists in the sphere of agricultural and horticultural instrumentation must include the following:

(i) Electronics instruments may all look very much alike but the apparent similarity may be superficial. The features which make one circuit more reliable than another has something to do with the design (which includes software, in the case of computer-based systems), the quality of the components used and the standards of assembly and testing by the manu- facturer. The thorough manufacturer will put equipment through 'burn-in' tests, to weed out the unreliable units, before putting it on the market. Careful design and testing of prototypes (hardware and software), testing of production units and scrapping of the unreliable ones, and thorough documentation and reliable after-sales service all cost money. The purchaser who would minimise risk must be prepared to pay for all of this.

(ii) Electronics equipment — like any other equipment — repays care and attention throughout its working life. In particular, cables and connectors should be checked from time to time; so should the batteries in portable equipment. Electro-mechanical equipment, such as printers, may need regular servicing. Above all, measuring instruments need to be checked for accuracy from time to time, either by submitting them to a calibrations service or by use of whatever on-site methods are available. Objects whose weights are known to a sufficient degree of accuracy can usually be found for checks on weighing equipment, for example, and containers of known volume may also be available. The use of domestic weights and measures for these purposes requires (a) domestic permission and (b) absence of health hazard.

(iii) Potential users who have had little or no experience with this type of equipment should consult specialist advisers if they can before taking decisions. In the agricultural and horticul-tural sector most advisory services nowadays have specialists in instrumentation and control. Investment in electronic equipment can be very profitable if it does the job required. Equally, the wrong equipment or equipment wrongly used can be a hindrance and a waste of money.

1.7 Further reading

General
(1981). Electrical and Electronics Reference Issue. *Machine Design*, **53**, No. 11.
Fink, D.G. *ed.* (1975). *Electronics Engineers' Handbook*. New York: McGraw-Hill.

Section 1.2
B.S.2643 (1955). *Glossary of Terms Relating to the Performance of Measuring Instruments*. London: British Standards Institution.
Topping, J. (1977). *Errors of Observation and Their Treatment. The Institute of Physics and the Physical Society Monographs for Students*. 4th edn. London: Chapman & Hall.

Section 1.3
Collett, C.V. and Hope, A.D. (1974). *Engineering Measurements*.

London: Pitman Publishing Ltd.

Hayward, A.T.J. (1979). *Flowmeters: a Basic Guide and Sourcebook for Users*. London: Macmillan. New York: Halsted Press.

(1979) *Proceedings of Conference on Weighing and Force Measurement, 'Weightech 79'*. London: Institute of Measurement and Control.

Weaving, G.S. and Filshie, J. (1977). 'A solarimeter utilizing silicon semiconductor diodes'. *Journal of Agricultural Engineering Research* **22**, 113-126.

Woolvet, G.A. (1977). *Transducers in Digital Systems*. Stevenage, Herts: Peter Peregrinus Ltd., on behalf of the Institution of Electrical Engineers.

Section 1.4

Carrick, A. (1979). *Computers and Instrumentation*. London: Heyden & Son Ltd. Philadelphia: Heyden.

Clayton, G.B. (1979). *Operational Amplifiers*. 2nd edn. Sevenoaks, Kent: Butterworth & Co. (Publishers) Ltd.

Meiksin, Z.H. and Thackray, P.C. (1980). *Electronic Design with Off-the-shelf Integrated Circuits*. New York: Parker Publishing Co. Inc.

2 Microelectronics and the Microprocessor

2.1 The development of integrated circuits

The foundation of modern electronics for measurement, data processing and control was laid with the invention of the transistor in the 1940s. Semiconductor devices were in wide use before then, notably several forms of diode and photocell, but the transistor could amplify a signal — a task hitherto requiring thermionic valves — with the result that semi-conductor equivalents of valve amplifiers, oscillators, demodulators, etc., started to take over in industry. In addition to these analogue circuits, transistorised digital circuits began to appear. The electronic digital computer — another development of the 1940s — was based on large numbers of electronic switches, providing the capability for binary arithmetic (all '0's and '1's) as well as the logic operations which direct the flow of data through the computer. The development of 'solid-state' counters, timers, switches (flip-flops) and multiple-input logic gates soon led to their uptake for industrial control systems, but in many cases initially they did not have a clear economic advantage over established electrical relay systems for sequential control of a process or over analogue instrumentation and control systems. Computer control was only worth considering for large-scale, high-capital industrial processes.

The situation changed dramatically, as most people are aware, with the introduction of the integrated circuit in the 1960s, leading to LSI, i.e. large-scale integration of many circuits on a single silicon chip, in the 1970s. Mass production of calculator and timing chips brought low cost and assured performance to industrial and domestic users.

Once the move towards digital systems was under way designers began to develop solid-state circuits for conversion of analogue inputs from transducers into digital form. The development of low cost, high accuracy A/D converters, coupled with the concurrent development of multiplexing circuits, gave rise to the data logger,

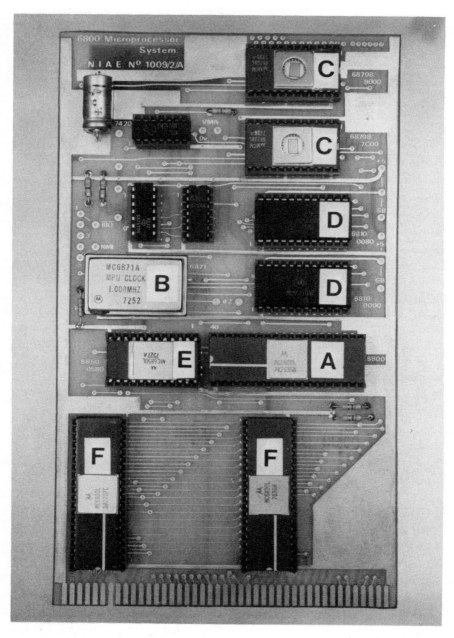

Plate 4 Single-card microcomputer: A. microprocessor, B. Clock generator, C. EPROMs (2 × 1024 byte), D. RAMs (2 × 128 byte), E. Serial data input/output, F. Parallel data input/output (NIAE).

which could sample many analogue inputs in turn automatically, at preselected times, displaying and recording the measured values along with those derived from any digital inputs in the system. Decimal digital displays were developed for these loggers and incorporated in DVMs (digital voltmeters). Standard ways of transmitting digital output data to typewriters or teletypes, punched paper tape and magnetic tape were adopted (appendix 2). Data on tape were transported to the nearest computer for final processing. The computer itself was gaining advantage from miniaturisation and lower costs of components, including the semiconductor memory with which the central processing unit (CPU) works, both for the programmed sequence of operations to be performed and for the binary arithmetic involved.

2.1.1 *The microprocessor*

The next, inevitable step in the development of integrated circuits was the production of the microprocessor, carrying a central processor on one chip. It was then possible to construct a microcomputer on a single printed-circuit card, as shown in plate 4. The microprocessor in the illustration is an eight-bit device, i.e. the instructions that it handles and the arithmetical data upon which it operates are processed in groups of eight binary digits; in other words, in single bytes. An eight-bit sequence allows 256 combinations of '0's and '1's (including 0000 0000), which somewhat restricts the arithmetic that the microprocessor can perform at each step in a programme. Nevertheless, the many eight-bit microcomputers in use testify to their range of applications. Some processors have fewer than eight bits, in fact. This applies to many in domestic appliances, for example.

Apart from the microprocessor chip, on the card there are two forms of semiconductor memory, on separate chips, each capable of storing the number of data bytes shown. Random access memory (RAM) is available for 'read' and 'write' operations during the setting up and execution of a computer programme: data can be stored at addresses (locations) in each RAM and recovered swiftly when needed. The EPROM (Erasable Programmable Read-only Memory), however, is only available for the 'read' operation. Its array of semiconductor switches has been set in a bit pattern which is available for reference but unalterable unless the whole programme is erased by subjection to strong ultraviolet light. The EPROM carries the programme which the microprocessor will follow. (Note: A non-erasable

PROM would do the same job.) The crystal-controlled system clock is the essential element for timing the operational steps of the computer. Three links are provided with the external world of transducers, displays and other input or output peripherals. Two parallel interface adapter chips act as the communication channels for data in parallel digital form (e.g. the output of an A/D converter or the input to a digital display). A serial link provides a bi-directional, serial binary data channel for communications lines. This simple system has been used in a variety of applications to agricultural instrumentation, each application involving microprocessor programme development equipment − and specialist programmers (see plate 5) − to set up the required bit settings in the EPROMs.

Plate 5 Microprocessor laboratory, with programme development and test facilities (NIAE).

The progress of microminiaturisation of circuits has continued since the introduction of the first generation of microprocessors. The 128 byte RAM in plate 5 was perfectly adequate for the applications for which it was designed but since then the 32 k byte RAM has become available and it has been predicted that further miniaturisation will make it possible to increase the density one hundred fold — before physical limitations are reached or, as seems more likely, manufacturing and testing problems become too severe. This amount of memory on a chip is sufficient to store the contents of a large book in digital form. Higher packing densities of microcircuits have made it possible to combine semiconductor memory and the microprocessor on the same chip, too, so providing an even more compact system for a range of the less demanding monitoring and control tasks. Through mass production the semiconductor industry continues to provide more on a chip for substantially the same cost.

Another outcome of the increase in packing density is that, if required, the microcomputer can be programmed in one of the higher level languages, such as FORTRAN, or BASIC, which is familiar to the users of personal computers. These languages are too prodigal of memory space for systems with only a few kilobytes of storage capacity but, given enough capacity, they enable the less specialised user to create and modify programmes.

The implications for industrial applications of computing are clear. In combination with more capacious bulk memory devices, with faster access to data at individual locations, powerful systems will be available at economic costs for individuals and small businesses. This distributed computer power will be available for automatic data logging, data processing and use of operational research models, for business management and forecasting. In the more limited context of instrumentation and control, the microprocessor will be able to do more of the signal processing than it does already, in combination with the transducers, op. amps., A/D converters and other devices described in chapter 1. For example, it will be able to refer to stored information on the relationships between measured quantities in far more detail and, as a result, to derive information on the state of an industrial process which is more useful for manual control of that process than a set of individual measurements could be. This increased capability to correlate input information will also extend the potential of microelectronics systems for automatic control of complex processes outside the chemical engineering industry, where computer control is well established. Operations such as linearisation, drift correction and filtering of data are part of the existing capability of microprocessor-based instrumentation.

2.2 Microcircuit devices

2.2.1 *Transistors*

Descriptions of microelectronics devices are full of specialist terms, related to their method of manufacture, their composition and the resultant characteristics. The type of transistor used in these circuits is of particular significance, so a few words on this subject may be of value to the non-specialist.

The action of the transistor depends on the implantation of impurities at particular points in a chip of refined material, e.g. silicon. By careful control of this process it is possible to create in the substrate material contiguous zones of p-type material (excess positive electronic charges) and n-type material (excess of negative charges). The npn transistor is the building block for 'bipolar' integrated circuits. This device has p-type material sandwiched between n-type material. The latter provides the 'emitter' and 'collector', which carry the main current through the device and the former supplies the 'base' which, like the triode grid (cf. 1.4.2), controls this current, in accordance with the applied emitter-base voltage. Another family of integrated circuits is based on the FET (field-effect transistor) in which the current between a 'source' and a 'drain' is controlled by voltage bias on a third, 'gate' electrode, which controls conduction through a 'channel' running from the source to the drain. The gate operates to deplete the channel conduction in some FETs and to enhance it in others.

Both bipolar and field-effect transistors are used in analogue integrated circuits (op. amps. in particular), the npn bipolar type being favoured for wide-band applications. Digital logic circuits and semiconductor memory cells also employ both types. Logic circuits (cf. 1.4.1) are of several classes, which include DTL (diode-transistor logic), TTL (transistor-transistor logic), ECL (emitter coupled logic) − all of which employ bipolar devices − and CMOS (complementary metal oxide semiconductor) logic, which employs a form of FET. DTL gates are convenient to make in integrated circuit form, and have other virtues, but TTL is more often used, since it has greater noise immunity and output current capability. ECL has an exceptional speed of response but it is less immune to noise and it needs more power than TTL. CMOS logic circuits are built from complementary FETs, i.e. those with n-channels and those with p-channels. These devices in combination produce integrated circuits which have very low power requirements and for this reason they are particularly suitable for application to portable, battery-powered equipment. Their packing density can also be much higher than bipolar devices but the latter have the advantage in speed of operation.

Similar comparisons apply to the bipolar and MOS transistor flipflops which are the basic units for semiconductor memories and to the shift registers which carry out serial and serial/parallel data transfers in computers (cf. 1.4.1).

2.2.2 *Hybrids and silicon sensors*

Hybrid integrated circuits   The devices discussed in the preceding sub-section are of monolithic construction, i.e. all the transistors and their associated passive components are implanted in the silicon substrate which forms the body of each chip. In contrast, some larger, 'hybrid' circuits are formed by depositing circuit patterns onto electrically insulating substrates, on which they may be linked with discrete components and monolithic circuits. The deposition techniques include evaporation of metal onto the substrate under vacuum, electroplating and electrochemical formation of oxide surfaces on a metal (anodisation). Silk-screen printing methods are widely used for depositing thick-film resistors. The construction of the humidity sensor shown in fig. 1.4 and the solarimeter diode array of fig. 1.5 provide two applications of these deposition techniques, though on a macro scale.

Hybrid integrated circuits of small size are particularly appropriate to the microwave region, over a frequency range of about 1 to 15 GHz. At these frequencies microstrip transmission lines can be made by photolithography etching and screening methods and small passive elements can be built into the dielectric substrate. This frequency band is used for Doppler radar speed meters (3.3.1), which can therefore use MICs (microwave integrated circuits) in combination with two small, inexpensive semiconductor devices, namely the Gunn diode and the Schottky barrier diode. The former generates microwaves when energised at low voltage and the latter detects the return signal.

The above example shows one way in which hybrid construction can find application in the age of monolithic devices. The physical basis of microwave transmission and reception requires components of particular forms and dimensions and the hybrid system is better able to accommodate them.

Silicon sensors   Many of the transducers described in 1.3 are stated to be 'compatible' with microelectronics circuits. This implies that they are capable of operating from low-voltage power supplies (i.e. below 12 V) and that they provide an output signal which requires little conditioning to make it acceptable to analogue or digital semiconductor circuits. Silicon strain gauges, temperature sensors and photocells are among these devices. However, as silicon techno-logy develops, an increasing range of monolithic silicon sensors is becoming available for industrial measurements. These sensors, made in a similar way to LSI components, can be mounted, hybrid fashion, in integral transducer/signal conditioning packages.

Silicon pressure sensors have been used very widely in the military and aerospace fields because they are small, with good linearity and fast response, and can survive the high temperatures and accelera-tions experienced in this environment. The smallest are less than 1 mm in diameter (cf. 1.3.1). These piezo-resistive transducers are expected to find a vast market for engine monitoring in automobiles (fig. 2.1) and, by extension, they will be used in monitoring agricul-tural field machines. One manufacturer's combined sensor and signal-conditioning circuits on a chip, for large-scale, low-cost pro-duction, converts absolute pressure up to 1 bar (10^5 Pa) into a 5 V FRO signal with a linearity of $\pm$ 1%; repeatability and hysteresis $\pm$ 0.13% and a temperature coefficient of little more than 0.01%/°C. This performance is obtained from a package of approximate size 5 mm x 6.5 mm.

Another integrated sensor/signal conditioning chip measures acceleration, with a sensitivity of 0.2 mV/ms^{-2}, by sensing the change in capacitance produced by the infinitesimal movement of a tiny cantilever beam under the acceleration force.

The piezo-resistive response of silicon can also be employed in flow measurement. Alternatively, heat transfer from a silicon device can produce an analogue measurement of flow, and the same effect is used to detect liquid levels in an on/off mode.

In the sphere of analytical chemistry ion-selective assemblies, based on FETs with special gates, have been developed for monitor-ing of individual constituents of liquids, such as nutrient solutions (chapter 4). A modified version of the ISFET (ion-selective FET) is also used as a pH measurement electrode.

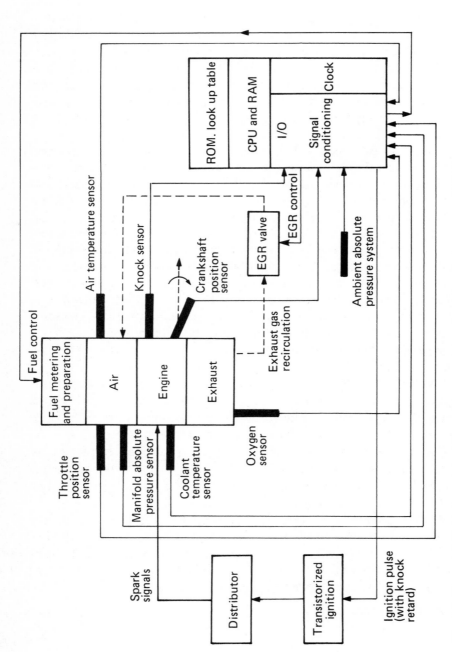

Fig. 2.1 The possibilities for automobile engine control, based on the microprocessor (IEEE).

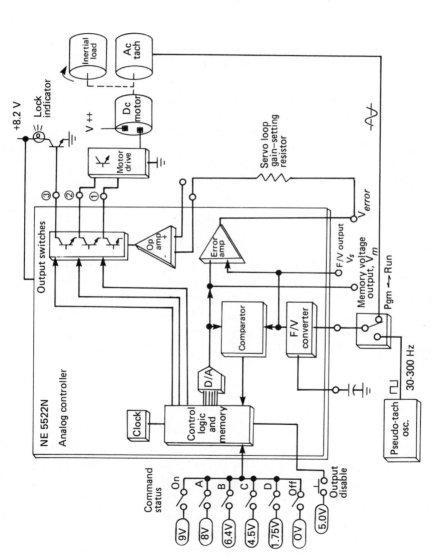

Fig. 2.2 Servocontrol chip for closed-loop control of a d.c. motor (Signetics Corp. and Machine Design Magazine).

2.2.3 *Analysis and control devices*

In the same sphere of analysis, the monolithic photolithography/ chemical etching process has been employed to fabricate a gas chromatography column, on a silicon substrate (cf. MIC devices) for analysis of gases. This sensor provides an analysis time of only a few seconds and has many potential applications in aerial pollution measurements.

A powerful analytical technique of another kind is the calculation of the correlation function between two fluctuating signals. This has many applications, including flow and speed measurement. The latter application is described in 3.3.1. Special LSI devices have been designed for this form of data processing.

Triac motor control elements were described in 1.4.2 and reference made to closed-loop control systems in 1.5. A link between the two is provided by the servocontrol chips which control start up and running of motors. These chips combine analogue and digital elements, as shown in fig. 2.2, which represents a system of closed-loop control for a d.c. motor, employing feedback from a coupled a.c. tachometer to maintain the desired speed during normal running. The digital section (control logic, memory and clock) can be programmed by manual commands and by simulated tachometer signals when the 'Pgm-Run' switch is in the Pgm (programme) position. The a.c. output from the real and pseudo tachometers is converted to a d.c. voltage by an F/V (frequency voltage) unit before comparison with the programmed value to determine the error at any moment (see fig. 1.7). The whole programming function can be taken over by a microprocessor, if required.

2.3 Complete systems

Before the application of microelectronics is discussed in relation to specific areas of agriculture and horticulture, in the ensuing chapters, some general points will be made, in part summarising what has gone before. Firstly, fig. 2.3 is a reminder of the elements of a comprehensive electronic monitoring system, with analogue and digital inputs and digital processing of the measured data, bulk storage and output to a display unit or controller. In particular, it serves to put the low cost of a microprocessor chip in perspective. This may be only a small fraction of the total hardware cost. Not all applications involve everything in fig. 2.3, of course, but some of the points in this section apply to systems with only one input and no bulk storage devices.

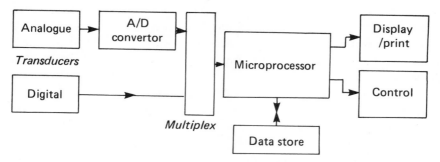

Fig. 2.3 The elements of a microprocessor-based monitoring and control system (NIAE).

2.3.1 *Equipment for field use*

Whether this equipment is hand portable or mounted on field machines and equipment it is likely to contain more than one input, although separate inputs may be selected by switch, rather than multiplexed; it is likely to have a LED or LCD (liquid crystal) display, and graphical presentation in the latter case; it is unlikely to contain a bulk storage device and it will almost certainly run from a battery, either of its own, or by connection to a vehicle's battery.

Sensors on field machinery may need fast response, to cope with the fluctuations in the measured quantity, if these are important. Multiplexing of such signals presents no problem with solid-state multiplexers provided that each channel is brought to the 1 V peak level or more (1.4.1). If low-level signals are involved the number of amplifiers required can be reduced to one by use of a reed-relay multiplexer, followed by the amplifier. Reed relays can handle signals down to the μV level but their relatively low switching rate limits their application to fluctuations of lower frequency.

Microelectronics equipment for field machines normally runs satisfactorily from a vehicle's battery if designed to cope with the otherwise catastrophic result of accidental reversal of the battery connections. Wiring from transducers is least liable to pick up electrical interference if the signals are digital. This also fits in with the move towards standardised electronic communications on field machines (chapter 3). Nonetheless many examples of trouble-free analogue systems can be found.

2.3.2 *Equipment for static installations*

The likely features of equipment for monitoring of buildings environment or processes within buildings, with or without control functions, are multi-point sensing, long cable runs and fairly slow change in measured variables. Bulk storage of measurement data is increasingly likely, too, as the power of small computers continues to grow. This leads to a greater likelihood that the equipment will be programmed to process the data into a form suitable for day-to-day management of the monitored operation.

The need to reduce the considerable costs of long multicore cables and the attendant risks of interference, including interaction between channels, argues for signal processing units near the sensing points, to reduce costs and interference effects (cf. section 1.6). More comprehensive systems, with additional local control functions, provide ideal applications for distributed processing.

The distributed processing system employs complete local monitoring and control units, each of which may include all the elements in fig. 2.3, except for the bulk storage device. These units are capable of performing independently, if need be, but they are connected to a central minicomputer, which holds the data for the whole system in bulk storage and which can command the satellite stations as required. The communications link can be a two-wire cable, carrying data in serial form. In areas of high electrical noise it can be replaced by a fibre-optics link, through opto-electronic couplers at each end.

In areas subject to mains-power interruption use of industrial standby power supplies will avoid loss of information and control. Similarly, standard commercial equipment is available to regulate mains-power supplies if their voltage varies significantly. In this connection, bubble memories have an advantage over semiconductor memories for data storage because the data in them are not affected by the loss of power (cf. 1.4.2).

Siting of this equipment is obviously important, both for access to it and to maximise its reliability. Transducers have to be where the measurements dictate but it is wise to find a clean, dry location for the main data processing equipment. Operators appreciate this, too.

2.3.3 *Software*

The particular point to be made here is the cost of development of the programmes developed for microprocessor systems. This can dwarf the hardware costs and must be borne by the user, one way or another. It is clearly worthwhile to use standard computer packages if possible in the same way that standard industrial hardware is cheaper and better tested than one-off designs.

Application of the mass-produced 'personal' computer to industrial monitoring, data processing and control provides a simple, if limited, system which the user is able to programme in 'user-friendly' BASIC routines for individual tasks but this does not seriously undermine the general point made above. Much depends upon the time and effort that the user is prepared to devote 'freely' to the programming task, the thoroughness of the programme testing and of the attendant documentation.

2.4 User training

In a fast-developing field anyone wishing to employ microelectronics devices is advised to maintain awareness of these developments, even if only in the most general way. Suitable literature abounds at all levels. For those with serious intent to learn more about these devices and their capabilities there is also a wide choice of short courses, providing 'hands-on' experience.

Manufacturers of agricultural and horticultural equipment who wish to incorporate microelectronic monitoring and control systems in their products can call upon a growing number of specialist companies which have experience in the design, production and servicing of electronics equipment for the farmer and grower.

2.5 Further reading

Aspinall, D. and Dagless, E.L. (1977). *Introduction to Micropro- cessors*. London: Pitman Publishing Ltd. New York: Academic Press Inc.

(1980). Electrical and Electronics Reference Issue. *Machine Design* 53 No. 11.

Fink, D.G. *ed.* (1975). *Electronics Engineers' Handbook*. New York: McGraw-Hill.

Gibson, G.A. and Lin, Y. (1980). *Microcomputers for Engineers and*

Scientists. Englewood Cliffs, New Jersey: Prentice-Hall Inc.

Hadley, L.J. (1981). 'New chips that simplify motor control'. *Machine Design* **53**, 109-112.

Meiksin, Z.H. and Thackray, P.C. (1980). *Electronic Design with Off-the-shelf Integrated Circuits*. New York: Parker Publishing Co. Inc.

Meindl, J.D. and Wise, J.D. *eds*. (1979). Solid-state Sensors, Actuators and Interface Electronics. *Institute of Electrical and Electronics Engineers. Transactions on Electronic Devices* **ED-26 No. 12**.

Osborne, A. (1980). *An Introduction to Microcomputers Volume 1. Basic Concepts*. Berkeley, California: Osborne/McGraw-Hill.

Zaks, R. (1980). *Microprocessors from Chips to Systems*. 3rd edn. Berkeley, California: Sybex Inc.

3 Field Crops

3.1 Introduction

This chapter is concerned with the present and likely future applications of microelectronics devices to monitoring of field operations in agricultural and horticultural production and to the control of the machines and equipment which perform field tasks. The definition of a field crop is taken to include grass, whether permanent or not, other forage crops and fruit.

The opportunities for electronic equipment to contribute to the timeliness and quality of field operations are many and varied. Monitoring of field conditions can benefit the farmer and grower; monitoring of machines and implements can help the operators of field machinery in their tasks, and automatic control can take over tasks that are difficult for men to do well or those which are both lengthy and routine.

By the early 1980s electronic equipment has established a significant foothold in this sector of agriculture and horticulture, with control of crop spraying a growth area. Some monitoring equipment has progressed from the add-on option to the standard feature, particularly on larger field machines, because when these are employed the cost of sub-optimal working or down-time through avoidable damage is exceptionally high. In consequence, this equipment is in quantity production. Conditions appear set for further expansion and serious consideration is being given to standardisation of connections between electronic units on tractors and implements, in anticipation of greater diversity of types and makes and the need for some degree of compatibility between them.

The two following sections are concerned with the development of electronics for machines and equipment, taking the latter first because, in the past at least, electronic monitoring of agricultural operations has been more widespread than monitoring of vehicle power units and ground drive systems. Monitoring of grain/straw separation in cereal combine harvesters has been included in the equipment section. The section on vehicle monitoring and control covers automatic guidance. Although the demonstrations of guidance systems over the years have not led very far, this remains a potential area for future developments. The ergonomics features of field vehicles are also touched upon, because their improvement — important in relation to operator safety and performance — also provides applications for electronics systems.

The next two sections deal with measurements of crop and soil properties, soil water and the weather, with their bearing on yield and the timing of field operations. The subject of soil water brings in considerations of monitoring and control of irrigation and drainage systems, while any discussion on the weather and on estimation of crop growth must include reference to satellite monitoring, which is beginning to make a contribution in the agricultural context.

3.2 Implement monitoring and control

3.2.1 *Draught, position and torque control*

Draught and position   Control of implement draught and position via the tractor's three-point hitch has been the preserve of mechanical sensors and hydraulic controls for so long that it might seem unnecessary to introduce electronics where a tried and trusted method exists. However, mechanical linkages are subject to friction, hysteresis and other causes of inaccuracy, including wear. The added problem of accommodating linkages in high power tractors with fully protective cabs has therefore led designers to adopt electrohydraulic control systems for large machines. This introduces simple electrical sensors with low friction and hysteresis, thereby improving the accuracy of control. It also provides flexibility in component layout and, through the use of electrical links between components inside and outside the cab, facilitates design of cabs for improved protection of the driver.

Figure 3.1 shows the basic design of a commercial electrohydraulic system. It contains three L.V.D.T. sensors (1.3.1) — two to measure draught force on the lower links and one to measure the position of the lift shaft. The movement of the lower links is restrained by a bar spring on each, with mechanical stops to prevent excessive deflections and the lower L.V.D.T.s have a stroke of ± 2 mm. The upper L.V.D.T. is compressed up to 10 mm by the movement of a cam on one of the lift shafts. The cam has the form of an Archimedes

spiral, giving a constant displacement per angular degree of movement. The control loop is closed by the hydraulic control unit and
the two cylinders which raise and lower the lift shafts.

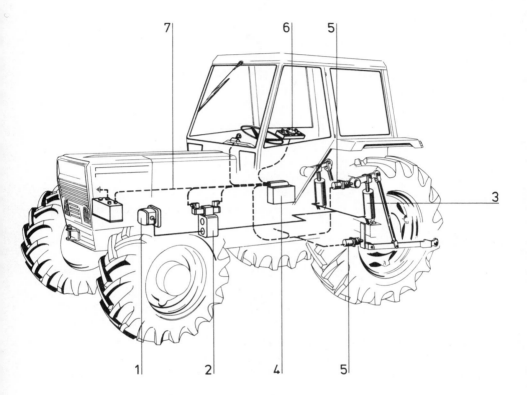

Fig. 3.1 Electro-hydraulic hitch control for agricultural tractors: 1. Hydraulic
pump 2. Hydraulic control valve 3. Hydraulic cylinder 4. Electronics unit 5. Lift
shaft and draught sensors 6. Operating panel 7. Cable harness (Robert Bosch,
GmbH)

This system demonstrates the greater flexibility of electrohydraulic
control, through its facility to mix the outputs of the three sensors,
under the control of an electronic unit, with an operating panel in
the cab (fig. 3.2). A variable proportion of the upper and lower
sensor signals is fed to one input of an operational amplifier, while
input settings from the operating panel are fed to the other. The
error signal drives the hydraulic control valve and hence the hydraulic
cylinders and the lift shaft. In the draught control mode the output
of the lower sensors is compared with the manual setting, while

in the position control mode the system responds to the output of the upper sensor only. Mixed control reduces the range of depth change brought about by responses to the signals from the draught sensors.

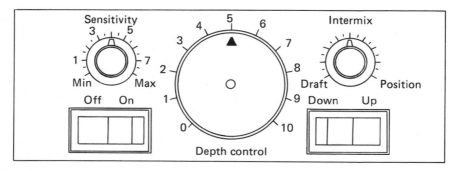

Fig. 3.2 Electro-hydraulic hitch control for agricultural tractors. Operating panel (Robert Bosch, GmbH)

The operating panel has a single input setting control, graduated 0 to 10, for any setting of the MIX control on its right in fig. 3.2. In the 100% draught mode LOWER (0 on the scale) is the position for greatest traction and RAISE (10 on the scale) least traction. The control valve is of the solenoid-operated type, with LOWER, NEUTRAL and RAISE positions and the on/off actions of the RAISE and LOWER solenoids are separated by a deadband which is adjustable in draught and position control terms by the electronic SENSITIVITY control. The system is rate-limited to reduce jerking of the implement and tractor. The DOWN limit of the input setting dial leaves the attached implement free to float. The UP/DOWN switch on the lower, right part of the panel provides a fast lift to a position limit shut-off (UP) and lowering again to the preset condition (DOWN). Safety is a major consideration, of course, and if settings are changed while the tractor ignition is off the system will not respond when the ignition is turned on again, unless the hitch position matches the input setting.

Torque   A strain-gauged pto torquemeter for torque and power (torque X pto speed) measurement has been used extensively for field testing of implements driven from the tractor's power-take-off shaft. The transducer fits onto the splined shaft – and presents an identical shaft to the implement but displaced rearwards by the length of the unit. The transducer, 250 mm long and 130 mm in diameter (plate 6), has a strain-gauged torque tube coupling these two shafts. The bridge-connected gauges are energised by an external battery, transferring current through two silver-graphite brushes, running on conducting silver slip rings in the rotating torque tube assembly. Two more rings and brushes transfer the output current to the electronic meter which reads torque, speed and power. The brushes are fitted into the captive outer casing, which carries an LED/photo-cell unit for digital measurement of shaft speed and the input/output connector to the battery and meter circuit. This unit has a range of 3000 Nm and a maximum error due to non-linearity, hysteresis and repeatability of ± 0.2% FRO up to 3000 r/min. The slip ring assembly has long life but periodic servicing, combined with recalibration in a torque rig, is recommended. In very demanding conditions it is possible to employ a non-contacting method for the supply of power to and the extraction of signals from a strain-gauged torque tube. A short-range telemetry system employs frequency-modulated radio coupling between a ring aerial around the rotating shaft and a microelectronics package on the shaft. The latter collects power from the aerial and retransmits an f.m. signal carrying the measurement data.

Torque and speed metering in the above ways provides a means to sense driveline overloads in addition to monitoring of normal working. The signals so derived can be used to control variable-speed drivelines and to initiate rapid disengagement of the drive in the event of a major overload.

The general role of microelectronics in this area is to condition the transducer signals; to perform any calculations required (i.e. multiply torque by speed); to carry out any tests required (e.g. check that torque is within safe limits) and to feed display and control units accordingly.

3.2.2 Crop establishment

Drilling   Monitors for seed drills are widely used, to alert the tractor driver quickly to a blockage in any of the seed delivery tubes. The sensor, placed near the bottom of each tube, is normally of the photoelectric type. The photocell receives illumination from a light source (filament type or, more usual, an LED) on the opposite side of the tube and registers the interruption of this light caused by the passage of the falling seed. A monitor unit in the tractor cab contains a row of lamps, corresponding to the row of delivery tubes on the planter. These alert the driver to the existence of a malfunction and identify the row or rows concerned. They are augmented by an audible alarm.

Plate 6 Pto torquemeter for tractors (pto guard removed for clarity) (British Hovercraft Corporation)

Some equipment of this type goes further and provides information on both planting density and area covered at any time (plate 7). This requires an additional sensor to measure distance covered, which can be a pulse generator, operated from a ground wheel, or a radar velocity sensor (see 3.3.1, where the recommended method of calibration of these distance measuring systems is outlined). The actual row spacing being used must also be set on one of the instrument's rear controls and it will then provide the required information on area covered. The population count/acre is displayed digitally and operation of the ROW SELECTOR switch provides a regularly updated read-out of the population count in a selected row. In the SCAN setting of this switch a read-out is displayed for each row sequentially at intervals, allowing a rapid check of all the rows for significant differences. Again, the row-width selector switch must be at the right setting or the readout will be in error.

The design of the photoelectric sensor assembly for this monitor raises a familiar problem, namely variation in planter design entails the production of sensors specific to individual models. Fortunately, the sensors are capable of detecting a variety of grains. The POPULA-TION SELECTOR control is normally set for corn (maize), for which the display automatically gives a direct read-out of population/acre. With other seeds it is necessary to refer to the planter manual for the desired population and to adjust the implement as required. A short test run is made to determine the average seed spacing (by uncovering the planted seeds) then, from this value and the selected row spacing, a figure for the extrapolated population/acre is readily calculated. In further test runs, with the instrument monitoring row 1, the POPULATION SELECTOR control is adjusted until this figure is displayed by the instrument. The setting is then used for that seed and the SCAN mode will show the row-to-row variations occurring with that type of seed.

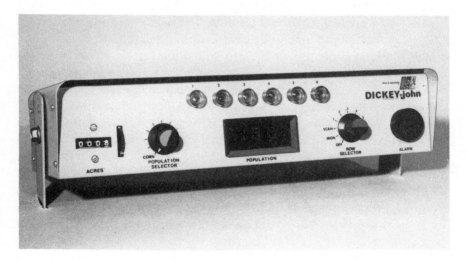

Plate 7   Seed planter monitor (Dickey-John Corporation)

An additional feature is the use of the array of lamps and the audible alarm on the monitor to indicate electrical fault conditions. Normally each lamp flashes as the photocell in the corresponding delivery tube registers the passage of the seed. In electrical fault conditions all the lamps will be off, while the alarm sounds. The alarm also sounds if a battery of the wrong voltage is connected.

Many farmers are now employing the 'tramlining' technique, leaving selected rows unplanted during seed drilling, to provide visible tracks in the growing crop as guidelines for the tractor in subsequent spraying and fertiliser applications. The bout widths of the sprayer and fertiliser distributer must be multiples of the drill width for this method to be effective, of course, and the sprayer may be several times wider than the drill. Therefore, in general there is no need to create tramlines on every drilling bout. A commercial electronic tramlining aid can be set to the required bout frequency and will prompt the driver audibly at the beginning of each tramlining bout by monitoring the lifting of the drill at the end of every bout. A manual 'inhibit' switch prevents interruption of the sequence by unscheduled stops and another manual control enables the driver to advance or retard its sequence at will, in order to fit in with particular areas of the field.

Transplanting   There has been a significant expansion in the types and numbers of crops established by growing seedlings in compost blocks and transplanting them semi-automatically or automatically in the field. Automatic machines take in long 'bandoliers' of seedlings, which the machine separates into individual units before planting them.

Although a high rate of germination and viable seedling growth characterise the block-growing methods empty blocks do occur and will be planted by automatic machines unless some detection and rejection method is incorporated. It is worth noting here that such methods were developed for an earlier, related task, namely the detection of seedlings for automatic rowcrop thinning. In good conditions two methods can be used successfully. One is based on a simple contact sensor, which forms part of the gate circuit of an FET. This probe is able to detect seedlings in the ground with a high degree of reliability, even when the ground is dry, owing to the high input impedance of the FET itself. The second and more complex method is non-contacting and can detect a seedling which is not upstanding from the soil surface to any great degree. In essence, the sensor employs a small lamp which illuminates a patch of soil, via a simple optical system. The same patch of soil is viewed by two silicon photocells in an integrated circuit. Both cells receive the reflected light from the soil via a dichroic mirror which directs the light in the waveband below 700 nm to one cell and transmits the remainder (the near infra-red light) to the second. An electronic circuit derives the ratio of the two output signals from the photocells. The reflectivity of soils varies considerably over the visible and infra-red wavebands and even more between soils of different types, yet the measured infra-red/visible reflectance ratio does not rise much above 5:1 except for peaty soils. On the other hand, if a seedling is in the field of view the chlorophyll in the leaves absorbs strongly in the green region of the visible band, with the result that the measured ratio is substantially greater than 5:1. Using this principle and scanning the light beam laterally to the direction of travel of an automatic thinner, it is possible to pick out brassica and lettuce seedlings growing in a range of soils, even at the peaty extreme. This sensor is therefore adaptable to the transplanter, to detect empty blocks.

3.2.3 Crop spraying

In the introduction to this chapter it was stated that the application of electronics to crop spraying is growing. This applies particularly to spraying of ground crops with hydraulic nozzles spaced out on booms. For this operation several similar electronic systems are available to the farmer and contract sprayer. They measure the forward speed of the tractor and the flow rate of the spray liquid from the bulk tank into the spray lines. The boom width and other details of the tractor and implement are entered manually into the electronic unit and this provides the driver with information on the work and application rates of the sprayer. In some cases the unit controls the flow rate according to the forward speed of the machine. These electronic systems, which will be described in more detail in the rest of this sub-section, are almost universally based on micro-electronics circuits and many employ microprocessors for their signal processing.

Speed measurement The simplest method in general use for measuring this quantity employs magnets fixed to a non-driven wheel and a sensor mounted on the wheel support where it will sense the passage of the magnets as the wheel rotates. A magnetically operated reed switch can be used as a sensor. Any method of counting wheel revolutions (or parts thereof) may be employed but the simple magnet sensor is robust and defies dusty conditions. The repetition rate of its output pulse is a measure of the rate of revolution of the wheel, of course, and this can be converted to forward speed if (a) the circumference of the wheel is known and (b) the wheel rolls over the ground without slip or skid. The quantity (a) is easily determined and is one of the numbers entered manually into the electronics unit if this method of speed measurement is used. As to (b), it is assumed that a non-driven wheel moves forward in a pure rolling mode and field tests have shown that this is so for most of the time.

The Doppler radar ground speed sensor is a more costly method but without the potential limitations of the wheel revolution monitor. It will be described in 3.3.1 and here it is only necessary to note that its signal is transmitted to the central electronics unit without the need for further signal processing.

Flow measurement   Corrosion-proof propeller flowmeters (1.3.3) provide one method of measurement in common use commercially. Like the wheel sensor they produce a digital signal which needs conversion to the measured quantity, in this case by comparison of their readings with measured quantities of liquid discharged, in a static test.

A second method is measurement by sprayline pressure. The quantity of fluid discharged is proportional to the square root of this excess pressure, as noted in 1.3.3.

Monitoring   The central display unit takes in the information mentioned above, and a manual setting of the implement's width. From this it can determine, and will display on command, such quantities as total area covered; incremental area covered from an initial point in time (say, since the spray tank was last filled); total quantity of spray applied; incremental quantity and application rate. Some also allow the setting of a target speed, fluid pressure or application rate and continuously display an analogue indication of the deviation from the target value, or operate out-of-limits lights, as aid to improved manual control of the task.

Control   Application rate should be linked to forward speed for uniform deposition of active material on the crop. Two ways of achieving this have been developed: one changes the flow rate to increase or decrease proportionately with speed and the other changes the dilution of the active material according to speed, while maintaining constant nozzle flow. The latter system is designed to avoid the wide range of flow rate through the nozzles inherent in the former method. Figure 3.3 outlines a closed-loop spray control system of the former type, with inputs from a pressure sensor and radar speed meter, and an output to a servo control valve in the

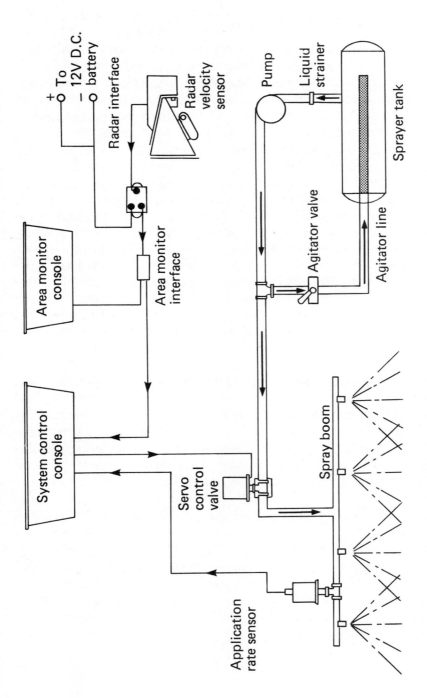

Fig. 3.3 Spray control system (Dickey-John Corporation)

spray supply line.

Another aspect of control is avoidance of under- or over-lapping of sprayed swaths arising from inaccuracy of steering by the tractor driver. This problem is more acute with wider booms. Although visible markers for the tractor wheels can be provided in several ways — for example, tramlining — these are not appropriate in all circumstances and other forms of bout width marking have been developed. One of these is the line of foam blobs, dropped from the tip of a spray boom, which is clearly visible and sufficiently long-lived to provide a guideline for the driver on the return bout. Unfortunately, it is still difficult for him to match the boom-tip to the line of the foam markers at the distances involved. One solution to this problem is provided by a linear array of photodetectors mounted parallel to the boom, at its extremity. Each detector has its own lens and views part of the ground beneath the boom tip, the whole array scanning about 0.5 m. The output signals are conveyed by cable to the tractor cab, where signal processing sorts out the position of the foam blobs relative to the photodetector array and indicates this position on a corresponding array of lights placed in front of the driver. With this steering aid the driver can hope to keep under or overlap down to ± 1%. The subject of tractor guidance is dealt with more generally in 3.3.2.

Calibration   Calibration of speed meters is covered in 3.3.1 and flowmeters can be checked in a static test, using collecting jars. There are also hand-held, calibrated flowmeters which can be quickly coupled to individual nozzles in turn to measure their output rate. The propeller sensor is coupled to a meter which registers flow rate digitally. These nozzle calibration units, which should be used regularly, are within ± 2% accuracy.

Boom suspensions   Boom flexing and swaying is inimical to uniform applications of expensive sprays, therefore designers constantly seek to reduce these forms of boom movement. Considerable advantage has been derived from the introduction of passive, sprung and damped suspensions but further advance may require active stabilised suspensions, which will involve electrical sensing of displacement and/or acceleration and the use of electronic servo control systems.

Distribution of granular materials This may be controlled by means similar to those just described, substituting a metering dispenser for the liquid flowmeter.

3.2.4 *Crop harvesting*

Grain Many types of self-propelled harvester are fitted with sensors which monitor shaft speed, loads on elements of the system, bin fill, and so on. They provide indications of malfunction or of the need for attention, through visual or audible warnings in the driver's cabin. Grain combine harvesters also employ loss monitors which sense the amount of grain at crucial points in the system. Commonly, acoustic sensors which are 'tuned' to respond preferentially to the impact of the grain are placed behind the sieves and the straw walkers. Sometimes a third sensor is placed in the machine's airstream, too. The acoustic sensors provide an electrical output, in the manner of a microphone, and the driver is given an analogue indication of the level of grain impacts at the monitoring points by a pulse-rate meter (plate 8). The sieve and straw walker sensors provide a measure of grain loss rate, which can be compared with the limit of acceptable loss decided by the driver. If losses exceed this limit the meter indication may be supplemented by a visual and audible warning that the driver should reduce the forward speed of the harvester. The sensor in the airstream provides additional information on combine adjustment.

One commercial loss monitoring instrument also has an input for a forward speed signal, which enables it to provide a loss/unit area indication. This puts the information into its most useful form. Automatic forward speed control is provided by another commercial instrument which employs an electronic tachometer to monitor engine speed. Forward speed is reduced automatically when a preset level of acceptable grain loss is exceeded, or when the engine r/min indicates the presence of overload.

It should be noted that these monitors provide an indication of loss rates, rather than a numerical measurement: for this reason their meters have arbitrary scales. The upper limit of loss must be set by calibration in the field.

Another aspect of combine monitoring is measurement of the mass of grain harvested from a particular area. It is possible to gather this valuable management information by attaching a compact, high-throughput mass flowmeter to the discharge spout of the combine. Each tankload, or part tankload, can then be weighed as it is propelled by the combine's discharge auger through the spout into an adjacent trailer. The attached transducer has to be compact in height, to avoid fouling the side of the trailer.

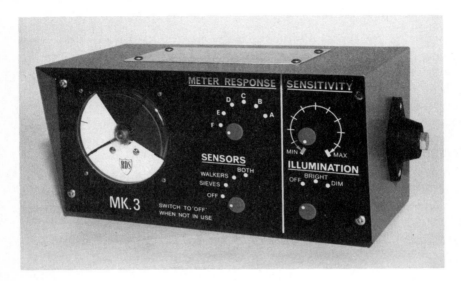

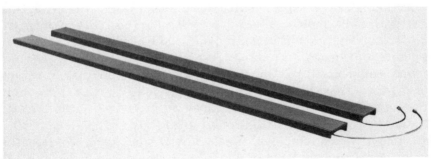

Plate 8 (a) Combine loss monitor (b) Acoustic sensors (RDS Farm Electronics Ltd.)

The outline and dimensions of one meter of this kind are shown in fig. 3.4(a). The larger shaded sections represent the framework by which the meter is attached to the discharge auger's casing. A metal cylinder, 420 mm in diameter and 288 mm high, is attached to the framework and contains an internal cylinder, 252 mm in diameter and 100 mm high, which is fitted with a conical cover, with 150° apex angle. The inner cylinder, which is attached to the outer one by radial struts (the smaller shaded elements), acts as a baffle for grain leaving the discharge spout and falling through the outer cylinder. The annular curtain of grain formed in this way falls onto the 60° slope of an inverted, open-ended cone, with the same diameter as the baffle. This funnel is attached to the main frame by three canti-lever arms, fitted with strain gauges which give an output dependent on the mass flowrate of grain through the funnel. Two of the arms are shown in fig. 3.4(a). The form of the associated electronic circuit is shown in fig. 3.4(b). When used with wheat or barley this device is capable of dealing with a flowrate of 100 t/h.

(a)

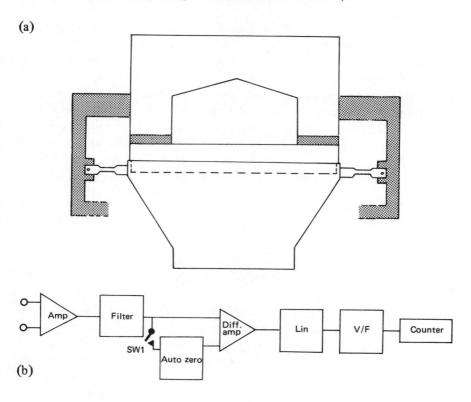

(b)

Fig. 3.4 Combine discharge meter (a) Sensor, (b) Electronics (NIAE).

Referring to fig. 3.4(b), the initial strain gauge amplifier is followed by a low-pass active filter which heavily attenuates any component in the signal above 0.4 Hz. This removes vibration signals generated by the combine itself. To eliminate transducer and amplifier drift, an automatic zero circuit tracks the signal from the filter. Its output is frozen when weighing begins and is fed to the non-inverting input of an operational amplifier during grain discharge. The next stage (linearity compensation) is needed because the output of the strain-gauged sensor is not linear with mass-flow, due to the differing interactions between individual grains and the funnel at high and low throughputs (there is more individual impact at low rates). The instrument's reading is also sensitive to grain type and moisture content, and to tilt from the vertical, so better than ± 5% accuracy is not to be expected of it. Greater accuracy can be obtained from it in static installations, where its compact shape makes it convenient as an insertion meter in handling systems for cereal grain and other granular materials. It must be calibrated for each type of material, however, and this is most conveniently done at a centre with facilities for circulating granular materials at controlled and known rates.

This discharge meter has been described at some length because it exemplifies several of the problems associated with the development of instrumentation for field use in agriculture. The common problems of non-linearity and drift apart, siting and space limitations, vibration effects and the effect of moisture content and type of crop, all influence the accuracy attainable and have to be taken into account in both design and use if even a modest accuracy, such as ± 5%, is to be attained.

Potatoes   An X-ray separator of potatoes from stones and clods was one of the earliest successful applications of electronics to field harvesting. Although its circuit design largely predated the microelectronics era, the X-ray sorting technique – now well-established – makes full use of microelectronics devices. The separation principle remains the same and is summarised here.

The transmission of low energy ('soft') X-radiation can be used to discriminate between objects of similar size but different atomic constitution, such as crop material on the one hand and stones or clods on the other. The crop material is composed largely of hydrogen, carbon and oxygen, all of which are of lower atomic number than silicon, in this context the main constituent of stones and clods. The differential absorption between silicon and the other elements named increases with the softness of the X-radiation but there is a lower limit in practice, because the absorption of X-rays and gamma rays (cf. appendix 1) by matter is exponential in form. Mathematically, if I_0 is the intensity of radiation falling on an object composed of, say, four materials and with a thickness x, the transmitted radiation

$$I = I_0 \exp\left[-(\mu_1\ell_1r_1 + \ldots + \mu_4\ell_4r_4)\, x\right]$$

where μ is the mass absorption coefficient for a specific material and radiation energy, ρ is the density of the material and r is the proportion of that material in the radiated object. This equation could represent absorption by a potato covered with soil. The characteristic of μ is that it increases steeply at lower energies for all elements and the reduction of the radiation by absorption (I/I_0), then becomes correspondingly large. Therefore, if the emergent beam intensity, I, is to be big enough to detect conveniently, without the use of a very powerful source (i.e. I_0 very large), it is necessary to pick an X-ray energy of just sufficient softness to achieve the required discrimination. In the present instance this is the ability to distinguish large potatoes from the smallest stones or clods in a mixture of the three. The use of an X-ray tube with a target (anode) potential of 30 to 40 kV has been found to meet this specification.

The essence of the X-ray sorter for a potato harvester is shown diagrammatically in fig. 3.5. The X-ray tube radiates horizontal beams through a set of windows in its collimator assembly. The beams irradiate a corresponding array of X-ray sensors. The harvested potatoes, mixed with the stones and clods that survive pre-cleaning, is fed onto an automatically-levelled conveyor and reduced to a single layer by rotating rollers above the conveyor (not shown in the diagram). The objects all fall through the array of X-ray beams and are classified by the signal processing circuits which follow the sensors and operate pneumatically-operated fingers. These are depressed to allow stones and clods to continue their downward fall but remain in their normal angled position to deflect potatoes into a separate discharge channel.

The original system was designed to segregate a minimum of twenty objects per second. Its X-ray sensors were scintillator crystals, which convert X-ray photons into visible light. The light output was then detected by sensitive photomultiplier tubes. The photomultiplier current pulse represented the value of I in the preceding equation and this was compared with a reference pulse in an operational amplifier circuit. If the measured pulse was below the reference level the presence of a stone or clod was deduced and the circuit then operated the appropriate finger or fingers after a preset delay.

This system was capable of distinguishing between a 150 mm thick potato and a 25 mm thick piece of sandstone. In trials it separated potatoes from stones and clods with only a few per cent error, mainly among objects less than 50 mm in size. It is worthy of note that many of the development problems were centred on the means of presenting a steady flow of objects to the X-ray system. This is characteristic of automatic grading.

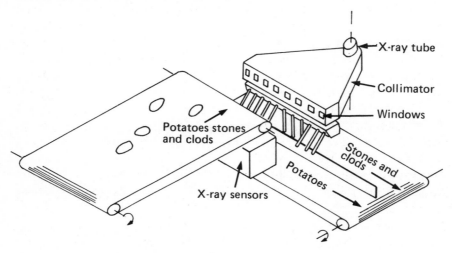

Fig. 3.5 X-ray sorter for potato harvesters (SIAE)

Transfer of potatoes from the harvester to a trailer or wagon can result in a significant amount of damage due to bruising. To reduce this damage a variety of boom height controllers have been devised. All seek to reduce the distance that the tuber falls before it meets either the bottom of the container or the level of the pile of potatoes already there. Some use mechanical sensors to detect the required level but non-contacting sensors, such as infra-red reflectance emitter/ receivers have been used successfully to operate electrohydraulic boom height controls.

Vegetable crops   Combines for peas and similar crops employ some of the monitoring aids found in grain combines but the special feature of vegetable harvesting, from an electronics standpoint, is the use of selective harvesting methods. Although mechanised once-over harvesting is understandably preferred to the additional costs of repeated visits to the field, manual selective harvesting endures in the horticultural sector and from time to time an auto- matic means of selective harvesting is devised. Here the X-ray absorption technique again finds application. For example, using radiation of the right degree of softness it is possible to distinguish between two lettuces of similar outward dimensions but very dissimilar bulk density, thereby making it possible to select lettuces of harvest quality by a non-damaging measurement of firmness. Silicon photodiodes have found use as X-ray sensors for this purpose.

Even the very hard gamma rays have been used on crisp-head lettuces in the field. This radiation would not be suitable for distinguishing between objects by virtue of their elemental composition, because μ varies little with atomic number at higher radiation energies: ℓ and x are the important factors. In the longer wavelength regions of the electromagnetic spectrum colour measurement has also been a basis for selective harvesting. However, whichever sensing method is used, the signal conditioning circuits are basically similar. The measured signal is compared with a reference signal in a discriminator circuit, which responds by accepting or rejecting each object presented to the sensor. Output signals feed electrically-controlled actuators of one kind or another.

Forage Tramp metal lying unnoticed in the field is a serious cause of damage and consequent delays to forage harvesting machinery during a crucial period. Smaller metal items may not cause damage to the machines but they present a hazard to livestock when the forage is fed out to them. Some commercial forage harvesters now incorporate an electronic metal detector, which detects any ferrous metal object entering the feedrolls behind the front pick-up reel and initiates almost instantaneous action to stop the feedroll drive. The sensor is mounted inside the lower feedroll (see fig. 3.6(a)), the latter being made of a material which transmits the magnetic field from two strong permanent magnets. To avoid any interaction with this field the upper roll is non-magnetic too. A figure-of-eight coil assembly surrounds the magnets and the whole sensor assembly is potted for dimensional stability. The sensor output is balanced unless a metal object entering the rolls upsets the symmetry of the arrangement: the moving parts of the machine do not destroy the symmetry and therefore do not trigger the sensor. When ferrous tramp metal is detected the output signal from the sensor is processed as shown in fig. 3.6(b). The filter section produces an output independent of the forward speed of the harvester and of electrical noise. This output is then rectified in the detector, for comparison with the threshold level. When this level is exceeded a solenoid immediately actuates a spring-loaded stop pawl on the feedroll drive. The main driveline is also put into neutral, to protect it from shock loads. The operator — alerted by the alarm signal — reverses the drive, to expose the tramp metal, and is then able to shut down the machinery while he removes the detected object. Interlocks prevent the harvester from operating until the metal detector is connected.

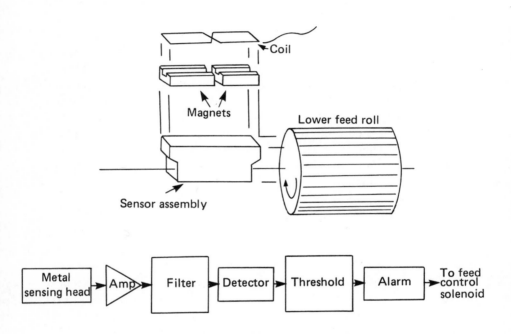

Fig. 3.6 Tramp metal detector for forage harvesters, (a) Metal detection sensor inside lower feed-roll, (b) Electronics (Sperry New Holland)

Another system, which reacts to any bulky objects picked up by the harvester, operates by detecting the impact of the object on the implement casing with a vibration sensor. Processing of the sensor signal is required in this case, too, to avoid false alarms.

A different aspect of forage harvesting is the application of pre-servatives to the crop as it is picked up in the field, to prevent deterioration in store. This treatment represents an additional cost on forage production but it may be unavoidable if the weather makes field drying too uncertain or prolonged, and it is likely to be no more costly than assisted drying in store. The amount of propionic acid or other preservative needed increases with the moisture content of the harvested crop and this introduces the need to measure forage moisture content. Methods of measuring this quantity are discussed in 3.4 but here it is appropriate to refer to the possibility of continuous moisture determination as the crop is picked up. 'On-line' measurement enables on-line control of the rate of application of the preservative, in proportion to the moisture content of the forage.

3.3 Tractors and other field machines

The preceding section was devoted largely to field implements, but not entirely so because reference to some of the functions of tractors and other machines was unavoidable. Similarly, although this section is about monitoring and control of tractors and other self-propelled field vehicles, the subject of implement monitoring and control cannot be kept out entirely because of the interactions that occur. However, it is possible to start with the machines themselves.

3.3.1 *Machine performance*

Monitoring equipment now available includes microprocessor-based units which display a range of information on the performance of a machine, including engine rpm and running hours; pto speed; the state of the battery; forward speed; wheel slip; area and distance covered in a set time. Each quantity can be selected by a switch and is displayed digitally. The first four in the above list employ conventional means of measurement and three are commonly found in vehicles of many types.

Forward speed Measurement of forward speed introduces the Doppler speed meter for 'true' ground speed, already mentioned in earlier sections. Some information on the way in which this works may be helpful.

Doppler meters for vehicles usually work in the microwave X-band, between about 9 and 11 GHz frequency. They employ Gunn diodes (2.2.2) to generate about 10 mW of power from a 6 V to 12 V battery supply and the microwave radiation is beamed from the unit by a horn of rectangular section. If this radiation (frequency f) is beamed downwards from a transmitter moving at a forward speed, V, with the axis of the beam at an angle θ to the horizontal, then the radiation reflected back from the ground to the transmitter will have suffered a Doppler frequency change, f, given by

$$f = 2 \text{ V F } \cos \theta / c \text{ Hz}$$

where c is the velocity of electromagnetic radiation in air, and is numerically 3×10^8 m/s. At a speed of 3 m/s, F = 9 GHz and θ = 30°

$$f = 156 \text{ Hz approximately}$$

Small frequency changes like this have to be detected in order to determine V. In addition there is not a single value of f in practice but a range of f $\pm$ Δf, because (a) the beam-spread produces a range of downward angles, θ $\pm$ $\Delta\theta$, and (b) surface roughness produces another — although much smaller — frequency spread. The reflected signal is picked up by the same horn and received by a 'mixer' diode (2.2.2) which also receives a small amount of the original signal from the Gunn diode. The resultant output is proportional to f and, following amplification, is displayed as forward speed. It is therefore important that the angle θ varies as little as possible from its calibration setting, which makes it desirable to mount the Doppler unit on a field machine as close as possible to the machine's centre of gravity, to reduce the effects of pitching.

The measurement scatter is less at small values of θ, but the return signal is also smaller and the angle chosen for θ usually lies between 35° and 45°. The reading is almost independent of surface moisture content for θ less than 45°.

On-farm checks on this equipment can be made by timed runs over a known distance at a selected meter reading.

Another non-contacting method for determining ground speed, without reference to wheels, driven or undriven, employs the cross-correlation technique (cf. 2.2.3). Essentially, the ground surface is viewed, vertically downwards, by two photodetectors spaced a little apart, in the direction of travel of a vehicle. The outputs from the two detectors vary in response to irregularities in the ground surface as the vehicle moves forward and are approximately the same but displaced in time by an amount that depends on the vehicle's V. The correlator takes in the two signals, subjects one to a progressive delay, then multiplies the undelayed signal by the delayed signal and averages the product. At a delay time T = d/V, where d is the distance between the two optical sensors, the output from the correlator displays a peak. Therefore, by measuring the delay T and knowing the distance d the speed V can be determined. This operation can be carried out digitally in several ways which differ in detail but, essentially, the analogue signals are first reduced to digital sequences by A/D conversion or by sampling their polarity (rather than their magnitude) repeatedly. Either way, one sequence is shifted relative to the other, one bit at a time, by a clock signal and the degree of correlation at each step is determined by summing the number of 0 and 1 coincidences. One commercial 64-bit, monolithic correlator chip provides a continuously up-dated 7-bit output (0 to 64) of the bit-for-bit agreements, and a threshold register containing a preset 7-bit number. The circuit produces a trigger signal when the output number exceeds the threshold number. In this way a cross-correlation peak can be detected. It needs external 'clocks' to control the sequence of operations and can carry out correlations at the high rate of 20 MHz.

Wheel slip In its simplest form the wheel slip of a tractor can be measured by comparing the speed of rotation of a driven (slipping) wheel with that of an undriven (no-slip) wheel, using the sensors already described, and making allowance for differences in wheel diameters. Where greater assurance is needed the speed of rotation of the driven wheel can be compared with ground speed measured by a Doppler meter or optical correlator. In fact, the speed of rotation of the driven wheel can be measured by viewing the periphery of the tyre with a Doppler or optical sensor and the two similar outputs compared, to generate a wheel-slip output. A dual Doppler system has been used in this way and found to measure percentage slip within ± 2%.

Distance and area covered The distance covered is derived by integrating the forward speed signal from one of the devices just described. Multiplication by the bout width then gives the area covered.

Fuel metering This measurement presents some problems in the case of diesel engines because not all of the oil passing along the supply line reaches the combustion chambers, i.e. the pressure relief flow and return should be measured. Nevertheless, fuel rate is measured via pipeline flow, commonly by monitoring the cycles of a piston or rotary metering-pump.

3.3.2 *Guidance systems*

Automatic steering systems developed over the years rely in some cases on cables and in others on contact between a mechanical feeler and furrow walls or crop material (e.g. fully-grown sugar beet or standing grain). In the latter type the movement of the feeler controls an hydraulically-powered steering system directly or, via simple electrical switches, through the action of an electro-hydraulic valve. Electronics has been employed in dead-reckoning systems (which tend to build up excessive errors), navigational beacon systems, energised leader cable systems and a non-contacting furrow-following system.

Beacons   The use of radio or optical beacons has often been considered as a means to fix the position of agricultural machines in the field. However, it is not easy for these systems to match the need to position a machine with ± 0.1 m precision (or even closer, for some operations) anywhere in a field. Commonly used radio navigation systems, based on fixed master/slave transmitters and a receiver on the vehicle depend on the precision with which the receiver can measure phase differences between the signals picked up from the transmitter at each point. This is a function of frequency. For example, a system working at 2MHz is unlikely to fix the vehicle's position to better than about 1 m. At higher frequencies this accuracy can be improved on but increasingly the system becomes dependent on line-of-sight operation, which limits its application to flat, clear areas. Line-of-sight limitations apply to optical beacons, too. A variant on the radio frequency system, devised for agricultural application, employs a single transmitter mounted on a field vehicle and a pair of spaced repeater units, which receive and retransmit the vehicle's radio signal. The return signals are processed by an on-board computer, to determine the position of the vehicle relative to the repeaters, from the phase delay between the transmitted and received signals. If the vehicle is off its programmed course, the driver is warned by an error signal, which he must cancel by a change of steering. The driver must also inform the computer when he has ended a bout, so that it moves on to its programme for the next bout.

Leader cables   The leader cable automatic guidance system has had the most success commercially of the electronic systems. It is an accurate system (± 25 mm) requiring the provision of a buried grid of insulated wires in the field, comprising a set of parallel wires, coupled together at each end, and a separate loop around the perimeter of the layout. The wire is laid below ploughing depth and fed by a.c. from headland, battery-powered circuits when the system is in use. The first commercial prototype worked with the parallel wires spaced at 18 ft intervals. A unit containing two search coils is mounted on a transverse boom at the front of the tractor and provides a phase-sensitive indication of its displacement from a central position over a buried wire. This error signal is used to steer the tractor into a zero error position. The search coil unit can be stepped along the front boom, thereby allowing the tractor to cover several successive bouts with guidance from the same wire. In this way an 18 ft spacing between wires can be subdivided into bouts of 3 ft, 4ft 6 in, 6 ft and 9 ft as well as 18 ft. The loop wire signal, received by the search coils, operates the on-tractor speed and implement controls, to allow the machine to turn at the headlands.

Although this system was designed for general field work it has found greatest use in orchards, where the restriction to following a fixed route year by year is perhaps less important. In this context it has been used for unmanned spraying, mowing, etc., day and night. For safety, the fuel supply is automatically cut off and the brakes applied if the front 'cow-catcher' guard or any of three emergency stop buttons are actuated, if the tractor veers off-course, develops prolonged wheelslip, loses its guidance signal or runs low on fuel, or if the cable energising box is illegally opened.

Non-contacting furrow following   Optical means have been devised for sensing the position of a furrow made by a mouldboard plough during primary cultivations or by a V-shaped tool set up specifically to create a guideline. A band of modulated light is projected onto the ground from a lamp unit mounted in front of a tractor. The band stretches laterally across the furrow and when viewed from a forward or rearward angle appears distorted by the changes of level introduced by the furrow. A viewing head, incorporating an array of semiconductor photocells, images the distorted light band onto the array. The associated electronic circuit determines the position of the furrow wall from analysis of the photocell signals. This information is employed to steer the tractor along the furrow via an electro-hydraulic servo system (fig. 3.7(a)). The projector and viewer are mounted on a transverse gantry, along which they can be tracked electrically, in order to be properly positioned for in-furrow or out-of-furrow operation.

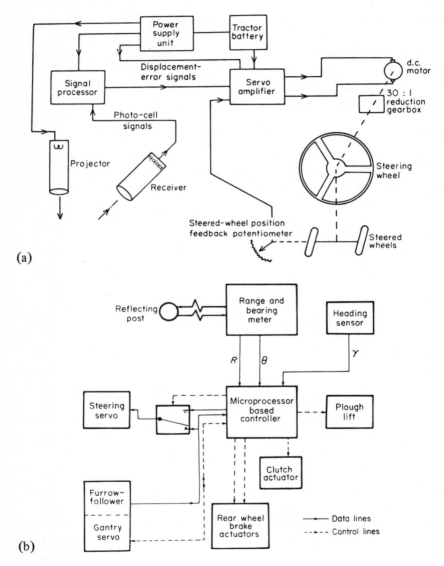

(a)

(b)

Fig. 3.7 Driverless, furrow following tractor (a) automatic steering system, (b) complete control system (NIAE)

With this system a tractor can be made to follow a furrow made under manual control and to repeat this operation on successive passes. It has been found to give comparable results to manual control when mouldboard ploughing.

To make the system fully automatic for driverless operation (excepting the first furrow and headland finishing) a means of automatic turning at the end of each bout must be employed. A solution is to provide reflecting posts at intervals (say, 10 m) along the headland and to mount an optical transmitter/receiver on the tractor, above the cab. A semiconductor infra-red diode lamp, modulated at a frequency in the MHz range, beams light vertically downwards onto a mirror, angled at 45° and rotating continuously about a vertical axis. This produces a horizontally scanning light beam which is reflected back to the transmitter when it strikes one of the boundary posts. The returned light is focused onto a semiconductor photocell and an electronics circuit measures the phase difference between the transmitted and reflected signals, which defines the range of the post at any moment. An angular position transducer on the mirror spindle provides the bearing of the post at the same time. From these data the distance ahead to the boundary can be calculated by an on-board microcomputer. At a preset distance from the boundary, the computer actuates a speed change, implement lift and then a programmed 180° turn, with lateral displacement. After that it manoeuvres the tractor and the furrow sensor on its gantry until the latter finds the last furrow, 'remembering' its original heading during the turn through integration of a rate gyro signal. The whole system is shown in fig. 3.7(b), which shows that the steering servo of fig. 3.7(a) is switched from control by the furrow-follower to programmed control by the microcomputer during the turn and then back to the follower when it has found the last furrow.

It will be seen that all the active parts of the system are on the tractor and that only passive reflectors are needed externally. The optical range and bearing system only comes into play at close range, so that there are no line-of-sight problems. There are also no constraints on the direction in which a field can be worked. Nevertheless, the system as outlined needs more sensors and circuits to monitor whatever agricultural operation it has been set to carry out and the U.K. Health and Safety Executive have laid down a set of safety requirements for automatic tractors, which inter alia, requires them to detect objects in their path and to stop immediately in this event or if there is any systems failure. Manual override is obligatory, too.

3.4 Crop monitoring

Farmers often need to measure the moisture content of a crop, to help them in timing harvesting operations, and frequently wish to estimate yield (or 'biomass') in the standing crop. Neither of these measurements is very straightforward but the measurement of moisture in grain has developed to an acceptable stage over many years and the grain moisture meter is one of the most widely used electronic instruments on farms. Moisture measurement is equally important in many of the post-harvest operations discussed in subsequent chapters but it is dealt with generally here, together with an appraisal of methods of yield assessment.

3.4.1 *Moisture measurement*

Electronic meters for grain moisture determination evolved in two directions, namely measurement of electrical conductance and measurement of the dielectric (capacitative) properties of moist grain, with calibrated ranges up to about 30% m.c.w.b. (moisture content, wet basis).

Resistance method   Resistance meters have a measurement cell with a pair of electrodes, commonly concentric rings in the base, which are bridged by the moist grain, under pressure. The electrical resistance (or conductance) of the bridging material is measured by an electronic meter circuit with a high input resistance, to cope with the drier material, which produces a cell resistance of about 100 MΩ. FET circuits are ideal for this application. This measurement depends more on the surface wetness or dryness of the grain than on its internal moisture so fine grinding of the grain is recommended for the most reliable results. Even so, for readings within $\pm$ 1% m.c. of oven reference methods the meter must be calibrated for each type of grain tested, and corrections must be made for ambient temperature, if this differs from the calibration temperature.

Capacitance method   Meters of this type usually contain a capacitance bridge circuit, using a two-electrode measurement cell as one arm of the bridge. The measurement is performed at frequencies in the 1 to 10 MHz range in order to limit the conductance effects produced by ionisable salts in the grain. Although it is unnecessary to grind the grain for this measurement it is necessary to obtain a standard bulk density of the sample in the cell, as far as possible. For this reason, the most accurate meters operate on a weighed sample, dumped into the cell in a standard manner, or weigh the sample after the cell has been completely filled and then make a correction for bulk density. The temperature of the grain must be measured, too, and a correction made for deviations from the calibration temperature. As with the resistance method, the calibration differs between types of grain and indeed within samples of the same type of grain. All calibration curves or tables represent mean values for a particular grain, averaged over years.

The need to refer to calibration tables and to make adjustments according to grain temperature and bulk density makes the signal processing in these meters an excellent application for the microprocessor, with attendant memory for storage of the tabulated data. Some newer commercial grain moisture meters incorporate a microprocessor for this purpose (plate 9).

Plate 9 Microprocessor-based moisture meter for eight types of seed, with averaging facilities. Right foreground: hopper for controlled filling of the sample cell, employing a spring-loaded slide. Capacity 100 to 180 g. (Sinar Agritec Ltd.)

Capacitance meters are also used for forage moisture determination, using large compression cells, but there are additional problems with forage crops. Unlike grain, the sample is often of mixed varieties and at various stages of maturity, therefore the calibration tables are averages of much more scattered data. However, there are other regions of the electromagnetic spectrum in which moisture measurements can be carried out with freedom from some of the interfering factors that occur in the MHz region.

Microwave absorption   In the GHz region measurements are little influenced by the presence of ionisable salts or by the distribution of water within the material. Here the Gunn diode finds another application, generating microwave power at around 10 GHz, to feed into a flat sensor containing a stripline or slotline. The signal reaching the far end of this line is detected by a diode whose output passes to the meter circuit. When forage, grain, or other moist material is packed against the sensor plate the attenuation of the output signal depends on the moisture content of the material, as well as its bulk density and, at these frequencies, the orientation of the individual pieces of the material. The orientation effect is especially pronounced in the case of strawy materials such as forage and the microwave stripline has to be made geometrically non-linear to offset this effect.

The microwave sensor is most convenient for measurements on baled or otherwise compacted material, since this provides a sample of reasonably constant bulk density. The measurement region extends well into the bulk of the material and, given repeated sampling, accuracy can be ± 4% m.c.w.b. or better.

Infra-red reflectance   This method is widely used in industry, particularly for on-line measurement of the moisture content of powdered and granular material. Its main limitation is that it measures moisture at or near the surface of the material and its accuracy is therefore dependent on the uniformity of moisture distribution in the material, as in the case of electrical resistance measurement. The infra-red reflectance meter irradiates the sample with light spanning two adjacent wavelengths in the infra-red region. At one of these water is strongly absorbing; at the other (the reference wavelength) it absorbs very little. The light reflected at the two wavelengths is measured by photodetectors with suitable filters and the ratio of the reflectance is a measure of the moisture content of the material, largely irrespective of the reflective properties of the material itself.

The traditional form of this meter is costly but developments in semiconductor technology have made it possible to design a much simpler, more compact and low cost instrument. This employs two low-power light-emitting diodes, each of which produces light output in a narrow waveband (about 50 nm) centred on a wavelength which can be fine-tuned by changing the current through it. One operates at a centre wavelength of 1300 nm, which is employed for optical communications along glass fibres. The other is specially manufactured to radiate on one of the adjacent water absorption bands. These tiny lamps, with integral lenses, are switched electronically at a high rate to irradiate a small area in turn. The reflected light is collected by a photocell, which 'sees' the light at the two wavelengths alternately and a following signal processing circuit extracts the required ratio.

Because the LEDs can be switched rapidly many sample readings can be taken per second and the circuit employs a microprocessor to display means and standard deviations after a selected number of readings, thereby providing well-defined values for the average moisture content of the material and for its variations.

The meter just described was developed particularly for forage moisture measurement, although its application is as general as that of the traditional design. It is capable of measuring forage moisture content to ± 3% m.c. One of its applications to this crop may be for on-line control of the addition of preservatives to the material at harvest, at a rate dependent on moisture content. It provides the rapid remote-sensing method of measurement needed for this task.

Although moisture meters are widely employed the moisture content of a crop is not the quantity which determines such important characteristics as keeping quality. The 'availability' of the water in the material is more important than its amount. This quantity, also known as the 'water activity', reflects the extent to which the water is free, rather than strongly bound to the crop material. Since the water activity in a crop is related to the crop's E.R.H. at a given temperature by the simple relationship E.R.H. = 100 (water activity) it can be determined by any method that measures E.R.H. A very convenient method is dewpoint measurement, which has already been described in outline (1.3.5). However, the specific application to crop moisture determination calls for a few more words here.

Dewpoint measurement   The meter shown in plate 3 can be used to find the E.R.H. of specific materials at known temperatures (and assuming one atmospheric pressure, 10^5 Pa) by placing the dewpoint sensor in an E.R.H. cell which is then packed with the material and sealed. The measured dewpoint and ambient temperatures at equilibrium allow the E.R.H. to be read from the scales on the analogue meter or taken from psychrometric charts or tables.

The relationship between E.R.H. (and water activity) and moisture content for a specific crop can be derived by measuring the E.R.H., as above, with samples whose moisture content is known from standard oven measurements. Once this relationship has been established for a range of E.R.H. and temperature the moisture content of a crop (stored grain, for example) can be measured by burying the sensor in the crop and allowing it to equilibrate. Since the sensor dissipates little power (about 0.2 W maximum) it does not affect local air conditions materially.

Overall, although the dewpoint method of measurement is considerably slower than many others it is an absolute method of humidity determination and it can be used successfully where others are difficult to apply. An example is measurement of the storage quality of oilseed rape. In terms of moisture content the upper limit is about 8%, which is low on the normal range of electrical moisture meters, whereas the corresponding E.R.H. is about 65%.

The advent of microprocessors and semiconductor memories makes it possible to increase the convenience of this method by storing the temperature/E.R.H./m.c. data for specific crops and automatically processing the measured temperatures to provide R.H. or m.c. on demand.

Measurement of the moisture stress in a standing crop is valuable on irrigated land, particularly when water is in short supply and must be used as sparingly as possible. Infra-red thermometers (1.3.7) provide a measure of this quantity, which is related to the difference between the temperature of the air and that of the vegetation. An infra-red thermometer directed at the vegetation provides a guide to its temperature, after due allowance for the leaf emissivity factor, which is about 0.95. The leaf-ambient temperature difference can be several degrees positive or negative, being negative (leaf cooler) if the plants are well supplied with water and positive if they are under stress.

3.4.2 Yield estimation

Electronic means of estimating forage biomass have been developed by many research workers. Capacitance measurement in the MHz region has been employed in most cases. Essentially, capacitance probes are placed in the crop and the moisture content of the enclosed or surrounding vegetation is measured. However, the variability of forage crops and of interfering factors such as dew has proved to be a serious limitation to the accuracy of this method.

In recent years, yield estimation has been one of the functions of remote-sensing apparatus (microelectronics-based, of course) carried by aircraft and satellites. The usual method of estimation is based on spectral reflectance, employing two adjacent wavebands, as in the infra-red moisture meter. In the present instance the two bands lie in the regions around 650 nm (the chlorophyll absorption band) and 850 nm. These wavebands have been used for estimating yields of forage and grain. The measurements clearly hold promise for large-scale surveys but they suffer from errors due to variations in ground cover by the crop, crop moisture content and the spectral quality of the ambient light.

3.5 Soil and weather

The crop environment in the root and aerial zones provides many potential applications for electronic monitoring and control. Few have been realised in farming practice, although research workers have many forms of instrumentation to call on. In consequence, this section is briefer than the importance of this topic merits.

3.5.1 *Soil properties*

Soil moisture, temperature, nutrient content and pH are all properties that are measured from time to time. Of these quantities, temperature can be measured by standard electronic thermometers and solutions prepared from soil samples can be tested for electrical conductivity and pH with equipment already outlined in 1.3.6. Soil moisture has been determined by a capacitance method, using an insulating probe with two stainless steel electrodes as an insertion sensor. The electrodes formed part of a high-frequency (30 MHz) circuit which provided an instantaneous output related to soil moisture content, with little dependence on soil type and temperature. The main problem with this and many other competing methods is that their response depends on the packing of the soil around the sensor. The infra-red reflectance method (3.4.1) does not suffer from this disadvantage, since it employs a non-contacting measurement; it may therefore develop as a rapid-response soil moisture meter, probably employing glass fibre 'light pipes' to convey the transmitted and reflected light between the meter and the soil, to avoid damage to the LEDs and photodetector.

3.5.2 *Irrigation and drainage*

Monitoring of the moisture status of root zone media is common-place in the protected cropping sector (chapter 4), and it provides the necessary input to automatic irrigation systems. In the field, automatic sequencing of pipeline irrigation can be found but closed loop control — based on soil moisture determination, evaporimeters (3.5.3) or plant moisture stress — is rare in farming practice. The problems of measuring representative conditions — or of multipoint sampling, with its attendant costs of multiple sensors and data links — are clear. The remote-sensing infra-red thermometer appears to have the greatest potential in this respect but its indications are subject to a variety of ambient influences which are not easily allowed for.

In the drainage sphere contractors can employ gradient-controlled ditch digging and pipe laying machines which rely on a laser beam to provide a gradient reference. The narrow beam is picked up by photodetectors, which adjust the share or mole-plough depth to maintain the beam centrally in the detector system.

3.5.3 *Meteorological measurements*

Electrical transducers are available for solar radiation, air tempera-ture and R.H., atmospheric pressure, wind, rain and evaporation. Most of these have been described in chapter 1. Rainfall can be measured by a version of the tipping-bucket weigher, which provides a count of the precipitation received, in units as small as 0.5 mm. An estimation of moisture loss from the field by evaporation and crop transpiration can be made from the amount of solar radiation received over a period but evaporimeters are sometimes employed. These allow free evaporation from a water surface or through a horizontal porous element and the water loss can be measured by a height gauge in a supply tank.

Electronics plays its part in provision of comprehensive, unattended data logging equipment, which collects the outputs from all the sensors, conditions and then processes them, to provide full records and on-the-spot information.

An interesting recent development is the use of compact radar sets to identify hailstorms, for attack by local hail dispersing equipment. On a lowlier plane, water sprays are employed as a means of frost protection for fruit buds in orchards, with some risk of damage to the buds and branches from excessive ice formation through needless spraying. This risk can be greatly reduced if the spray is triggered electronically by thermistor-based sensors placed at exposed points in the trees. The sensors can be designed to simulate the way in which fruit buds and flowers lose heat at low temperatures.

3.6 Future developments

The development of mass-produced, integrated sensor/circuit devices for production-line automobiles is gaining pace and will lead to digital monitoring and control of the engine along the lines shown in fig. 2.1. This trend is bound to affect tractors and other self-propelled field machines. These may well be required to run on more than one fuel by the end of the present century, thereby providing additional control tasks. Digital monitoring and control of transmissions, ground drives and mechanical and electrical power drives is a natural extension of this process, especially for larger machines.

Implement control, including the harvesting elements in self-propelled combines, is likely to move further towards electrohydraulic control, too. New sensors, such as the infra-red reflectance device for moisture measurement, will almost certainly extend on-line monitoring of soil and crop conditions for improved control of field operations, either manually or automatically.

The proliferation of electronics units on field machinery and implements — and the need for them to interact with each other at times — will inevitably lead to problems of compatibility between units from different manufacturers and between older and newer designs from the same manufacturer, as stated in the introduction to this chapter. The obvious technical solution to this problem is the adoption of standard couplings and communications links, preferably the international standards developed for data processing and computing (appendix 2).

Although extensions of automatic control are to be expected, and some field operations may employ automatic guidance of the machine, ergonomics and safety improvements to the driver's environment and task are likely. The display of information from the monitoring points on machines and implements requires attention from ergonomists.

Greater isolation of the driver from noise and vibration may require active electrohydraulic cab suspensions rather than passive springing and damping of seats. This greater isolation, however achieved, will strengthen the need for hazard and malfunction monitors among the operator's aids in the cab.

More powerful, computer-based equipment for local weather monitoring and forecasting should become available as the amount of information that a microcomputer can store and process increases. Forecasting, even in a limited way, requires considerable memory accommodation. Satellite monitors are already able to pin-point areas considerably less than 1 ha. Therefore they offer potential help on crop and soil monitoring to individual farmers, given the development of reliable methods of remote-sensing for these tasks. Improved meteorological information and forecasting clearly offers invaluable help to the farmer in his decision-making on drilling, spraying, irrigation and harvesting, in particular.

3.7 Further reading

General
Culpin, C. (1981). *Farm Machinery* 10th edn. St Albans, Herts.: Granada Publishing.

Section 3.2
3.2.1
Hobbs, J. and Hesse, H. (1980). *SAE Technical Paper Series 801018. Electronic/Hydraulic Hitch Control for Agricultural Tractors.* Warrendale, Pennsylvania: SAE Inc.
3.2.2
Cox, S.W.R. and McLean, K.A. (1969). 'Electrochemical thinning of sugar beet'. *Journal of Agricultural Engineering Research* 14, 322-343.
Hooper, A.W., Harries, G.O. and Ambler, B. (1976). 'A photo-electric sensor for distinguishing between plant material and soil'. *Journal of Agricultural Engineering Research* 21, 145-155.
3.2.3
Butterworth, H.M. and Butterworth, W.R. (1981). 'An overlap indicator for wide field machines'. *Transactions of the American Society of Agricultural Engineers* 24, 52-54.
3.2.4
Hooper, A.W. and Ambler, B. (1979). 'A combine harvester discharge

meter'. *Journal of Agricultural Engineering Research* **24**, 1-10.

Lenker, D.A. and Adrian, P.A. (1980). 'Field model of an X-ray system for selecting mature heads of crisphead lettuce'. *Transactions of the American Society of Agricultural Engineers* **23**, 14-19.

Palmer, J., Kitchenman, A.W., Milner, J.B., Moore, A.B. and Owen, G.M. (1973). 'Development of a field separator of potatoes from stones and clods by means of X-radiation'. *Journal of Agricultural Engineering Research* **18**, 293-300.

Palmer, J., Kitchenman, A.W., Moore, A.B. and Owen, G.M. (1973). 'Adaptation of a commercial potato harvester to a commercial X-ray separator'. *Ibid.*, 355-367.

Quick, G.R. *ed.* (1978). *ASAE Publication 1-78. Proceedings of First International Grain and Forage Harvesting Conference.* St Joseph, Michigan: American Society of Agricultural Engineers. Bohman, C.E. and Stiefvater, T.L. 'Electronic metal detection for forage harvesters', 270-273.

Theiss, W.A. and Garrett, R.E. (1974). 'Determining maturity distributions in crisphead lettuce fields'. *Transactions of the American Society of Agricultural Engineers* **17**, 1060-1063.

Section 3.3
3.3.1

Fritsche, R. and Mesch, F. (1973). 'Non-contact speed measurement – a comparison of optical systems'. *Measurement and Control* **6**, 293-300.

Stuchly, S.S., Thansandote, A., Mladek, J. and Townsend, J.S. (1978). 'A Doppler radar velocity meter for agricultural tractors'. *Institute of Electrical and Electronics Engineers. Transactions on Vehicular Technology* **VT-27**(1), 24-30.

Thansandote, A., Stuchly, S.S., Mladek, J. and Townsend, J.S. (1977). 'A new slip monitor for traction equipment'. *Transactions of the American Society of Agricultural Engineers* **20**, 851-856.

3.3.2

Burnside, C.D. (1971). *Electromagnetic Distance Measurement.* London: Crosby Lockwood & Co Ltd.

Harries, G.O. and Ambler, B. (1981). 'Automatic ploughing: a tractor guidance system using opto-electronic remote sensing techniques and a microprocessor-based controller'. *Journal of Agricultural Engineering Research* **26**, 33-53.

Morgan, K.E. *ed.* (1967). *Proceedings of Agricultural Engineering*

Symposium, Silsoe, Beds, Institution of Agricultural Engineers.
Finn-Kelcey, P. and Owen, V.M. 'Leader cable tractor guidance'.
2.18, 1-8.

Section 3.4
Hooper, A.W. (1980). 'Estimation of the moisture content of grass
from diffuse reflectance measurements at near infra-red wavelengths'. *Journal of Agricultural Engineering Research* **25**, 355-
366.
Matthews, J. (1963). 'The design of an electrical capacitance type
moisture meter for agricultural use'. *Journal of Agricultural
Engineering Research* **8**, 17-30.
Tucker, C.J. (1980). 'A critical review of remote sensing and other
methods for non-destructive estimation of standing crop biomass'.
Grass and Forage Science **35**, 177-182.

Section 3.5
Hamer, P.J.C. (1980). 'An automatic sprinkler system giving variable
irrigation rates matched to measured frost protection needs'.
Agricultural Meteorology **21**, 281-293.
Thomas, A.M. (1965). *ERA Report No. 5032. An Experimental and
Theoretical Study of In Situ Measurement of Moisture in Soil and
Similar Substances by 'Fringe' Capacitance.* Leatherhead, Surrey:
Electrical Research Association.

4 Protected Crops

4.1 Introduction

The protected crops sector is associated with a wide variety of vegetables and flowers. It is usually taken to include the mushroom industry, too. Vegetable and flower production — always 'under glass' at one time — is frequently under plastic of one kind or another now, much of which is in the form of tunnels, with both light and heat derived only from solar radiation.

This chapter is concerned with the smaller — but still very significant — area of heated greenhouses, clad with glass, film plastics or rigid, structured plastic materials. The cost of these structures, with all services and their running costs, are such that greenhouse growers have always been in the forefront of those ready to take up new technology if it appears to offer some improvement in efficiency. This quest for even greater efficiency has been heightened by the need to conserve increasingly costly fuel supplies (usually oil) which represent about 40% of production costs in northern Europe.

The search for cheaper energy supplies has led to the use of reject (waste) heat from industrial plant, and from electricity generating stations in particular. The use of solar, wind and geothermal power is also being actively investigated wherever local conditions favour these sources.

All of these factors have been responsible for the increasing application of computer-based monitoring and control in greenhouses, replacing traditional analogue controllers for the aerial environment and covering the root zone environment, too.

Rooting media have been seriously challenged in recent years by the form of hydroponics culture known as NFT or NFC (nutrient film technique or culture). This provides further applications of electronic measurement.

Mushrooms have their own special needs for aerial and compost monitoring and control. The introduction of electronics and computers has been slower in this sector, but the larger producers are moving towards the greenhouse industry in this respect.

In the ensuing sections of this chapter the aerial and root zones in greenhouses are discussed separately. Mushroom culture is dealt with next, before a section on possible future developments in the whole protected crops sector.

4.2 Greenhouse aerial environment

The heated greenhouse, whether of glass or plastic, is designed either for maximum light transmission (in northern latitudes light levels can have more economic importance than fuel costs) or for a combination of good light transmission with moderate thermal insulation where maximum light is not essential. If maximum light transmission is required the use of a double skin structure is disadvantageous, unless it can be built with an absolute minimum of opaque support members, to offset the unavoidable reduction in light transmission produced by reflection at both the outer and inner skins. Supplementary lighting from high efficiency discharge lamps (mercury or sodium, commonly) may be used at low light levels, while arrays of fluorescent tubes are employed as daylight substitutes in growing rooms, where seedlings are raised. However, this introduces additional costs, which must be justified by improved growth, uniformity and quality of the crop. In general, growers seek to use as much natural daylight as they can.

The single skin glass or plastic house is inevitably very responsive to changes in outside weather conditions and environmental control equipment has to contend with the resultant fluctuations of demand. It must perform well, too, because even ± 1°C non-uniformity of air temperature in the house can affect fuel costs and yield substantially. The double-clad house is less demanding in this respect but it is usually a less leaky structure and tends to present problems of humidity control.

Heat distribution (solar energy apart) is often by low-level pipes, with circulating hot water. Alternatively, warmed air may be discharged directly into the house from fan-driven convectors or distributed via perforated film plastic duct systems which, properly designed, give a better uniformity of temperature. Radiant heating of the crop, which might have energy advantages, is only in the experimental stage, commercially. Ventilation in glasshouses is normally either by hinged, motor-driven ridge and side ventilation panels, whereas plastics houses usually employ heavy-duty electric fans. Fans are also used to support air-inflated plastics structures.

Modification of day-length is required for control of flowering in some decorative plants (chrysanthemums, in particular). Motor-driven 'photoperiod' covers (usually of black polyethylene) are therefore installed. These are driven over and round the crop for the required periods, by motor. While they are drawn over the crop they also reduce heat loss, since they also cover the heat distribution system. This additional advantage has led to a wider use of unrolling and retracting covers for energy saving at night. These covers, now known as 'thermal screens', are used with many glasshouse crops, wherever they can be fitted into the structure of the house and the method of crop culture. In the latter context, tall crops such as tomatoes and cucumbers are often supported from roof trusses, therefore an additional internal support framework is required before screens can be employed. A second necessity is that in the withdrawn (furled) position screens must interfere as little as possible with light levels.

For some leaf crops (lettuce, for example) it is advantageous to increase the CO_2 level in the air from its normal level of about 350 v.p.m. (volumes/million) to around 1000 v.p.m. CO_2 enrichment is sometimes based on release of the compressed gas from cylinders but more often by burning propane.

It is essential that the gas combustion products from air heating and CO_2 enrichment equipment do not include sulphur, since this pollutant quickly damages plants even at very low concentrations. Pollution of the outer surface of the greenhouse has to be minimised, too, or light levels will be affected. The chief culprit in this latter instance is often the emission from the boiler installation on the greenhouse site. Acid deposits from the boiler are very difficult to remove from glass surfaces, except by etching with an hydrofluoric acid solution.

One other controllable feature of the aerial environment which must be mentioned here is the mist spray, which is sometimes used to create an atmosphere beneficial to the rooting of cuttings and which also finds application as an aid to fruit setting in tomatoes and other crops.

4.2.1 *Measurement transducers and meters*

Climatic   External climate sensors included several already described in chapter 1. Direct and diffuse sunlight are measured by solarimeters. A 'shade-ring' can be mounted over the sensor to measure the latter quantity separately but this information is of more concern to the research worker and designer than to the commercial grower. Air temperature is measured externally, as well as internally, by sensors in aspirated screens, to avoid radiation errors. Wind speed and direction involve cup anemometers and directional vanes with electrical outputs. Apart from its cooling effect on the exterior of the greenhouse, wind can affect the ventilation rate, particularly in glasshouses with ridge ventilators and, if strong enough, can threaten these parts of the structures. Therefore, in some winds it may be necessary to override the normal ventilation control and to shut the windward ventilators, at least. The onset of heavy rainfall is detected by an exposed electrical conductivity cell which can be 'reset' with an electrical heater when the rain diminishes or ceases. This, too, is needed for closure of ridge ventilators, to protect the crop inside.

Internally, apart from air temperature, humidity may be monitored, using one of the methods described in 1.3.5. Internal solar radiation may be measured, too, as a check on light transmission, or PAR sensors may be used to measure the radiation in the 'photosynthetically active radiation' band (400 nm to 700 nm). These employ blue-sensitive silicon photocells, covered by optical absorption filters of the required spectral range.

In principle, it should be better to control the temperature of the crop leaves than to control the air temperature in the house. This can be done with the infra-red thermometers already described, although they are far more expensive than simple thermistor or metal resistance sensors, and the emissivity correction must be applied (cf. 1.3.7).

Carbon dioxide CO_2 measurement introduces the infra-red gas analyser, in which the absorption of light by a column of air drawn from the house is measured photoelectrically at about 4 μm. A column of dry CO_2-free air is employed as a reference. The IRGA is an excellent instrument but costly. A less costly alternative is provided by an electrical conductivity method, originally developed to detect people buried in snow drifts. Gas drawn from the greenhouse is bubbled through a cell containing deionised water, in which CO_2 dissolves, creating weak carbonic acid. Electrodes in the cell measure the conductivity of this solution in the normal way. The water itself is continuously recycled through a column of deionising resin (plate 10).

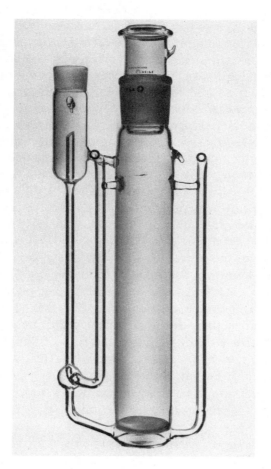

Plate 10 Continuous flow, conductimetric CO_2 sensor. Conductivity cell on left; column for deionising resin, centre (NIAE).

Heat meters Where greenhouses are heated from oil, gas or coal-fired boilers the energy input over a period is readily determined from the amount of fuel consumed during that time. If continuous measurements of heat energy are needed, in connection with the operation of thermal screens or for any other reason, there is no developed system comparable in convenience or accuracy to the use of a kWh meter in electrical heating systems. In commercial piped hot water installations heat input can be calculated with an accuracy between about 3% to 6%, depending on the heat loading of the system, by a combination of temperature and flow measurement. The difference in water temperature between the flow and return pipes is measured and multiplied by the water circulation rate; the product is then integrated to give an output proportional to the instantaneous heat supply. For highest accuracy, corrections should be made for variations in the density and specific heat of water with temperature.

The multiplication and integration of the temperature and flow rate is straightforward electronically, given the use of electrical resistance thermometers and flowmeters with an electrical output. Temperature corrections can be incorporated too, from stored data in a semiconductor memory. Accurate measurement of temperature is essential, however, since the difference between flow and return temperatures is usually small (between 5°C and 10°C). Flow measurement presents a different problem, arising from the impurities often found in the water in commercial heating systems. These may be in the form of magnetic particles which quickly foul flowmeters with rotating parts in the water flow or with magnetic couplings to external pulse-counting circuits. More expensive flowmeters are therefore needed. Meters of the venturi or orifice plate type (1.3.3), with differential pressure transducers, offer the best prospect of reliability, although some forms of vortex meter may prove acceptable. These present a 'bluff body' to the flow of liquid, which sheds vortices downstream in a regular manner when the flow rate is above a certain minimum. The pressure changes resulting from the passage of these vortices are measured electronically to provide a signal whose frequency is proportional to flow rate.

Fuel efficiency meters   Portable instruments are available for rapid checks on the efficiency of boilers. These are of special importance to greenhouse growers, who have both fuel costs and pollution to plants and structures to consider. One compact, microprocessor-based instrument of moderate cost employs a replaceable electrochemical sensor which measures the oxygen content of the flue gas, together with the temperature of the gas, which is drawn into the instrument through a probe by a pump. The user presses a button to indicate which type of fuel the boiler is burning (oil, solid, gas) before starting the sampling sequence. Within a minute the digital display on the meter will read, in turn, gas temperature, oxygen content and calculated fuel efficiency.

Feedback sensors Ventilators and proportioning valves in pipelines are fitted with displacement transducers (1.3.1) to indicate their position to the closed loop control equipment which operates the heating and ventilation systems.

Water sensors Mist spraying of cuttings and other plants, to maintain aerial moisture at a high but not disease-risking level, is usually based on electrical conductivity sensors which provide the input for electronic control of a solenoid valve in the water line. These sensors are quite simple devices which trap the spray and so change from high to low electrical resistance. Some have two electrodes embedded in a porous water absorber. Once triggered, the spray remains on for a preset time, after which the 'artificial leaf' (the sensor) dries out at a rate comparable to that of the surrounding plants until its resistance rises again to the triggering value.

Calibration Check of some transducers mentioned in this section can be made by the grower. Temperature sensors are unlikely to drift far from their original calibration but can be checked in melting ice fairly readily. Wind speed and direction indicators rarely give trouble, and any malfunction is likely to be self-evident. Standard salt solutions can be used for checks on humidity sensors (1.3.5). On the other hand, radiation sensors must be referred to a qualified laboratory for verification. Calibration gases of known pressure and CO_2 concentration are available for checks on CO_2 meters but this job is probably best left to the specialists also, as are maintenance and recalibration of water flowmeters.

4.2.2 Control systems

Heating and ventilation control systems of the analogue type used in greenhouses have employed screened temperature and humidity sensors to generate error signals in the normal way by comparing their outputs with the corresponding set points (cf. fig. 1.7). Signal processing has also been conventional, with an output dependent not only on the magnitude of the error (proportional control) but also on its integral and sometimes on its rate of change (2- and 3-term control). The output was passed on to the motorised valves and the ventilator motors which modify the heating and ventilation rates, respectively, and these in turn provided feedback signals to the controller. Under microprocessor control little has changed so far.

Programming can cause the digital controller to mimic three-term analogue control without difficulty but the greater range and flexibility of digital techniques still have to be exploited in the greenhouse context (see 4.5). The first step under digital control has been to incorporate additional features into the system — a task that would not be straightforward in an analogue system. The multiple input capability of the digital system has also made it simpler to sense temperature, humidity and other inputs in more than one place within the greenhouse. This provides much more information on the spatial and temporal variation of these quantities and it enables the controller to operate on polled averages of the inputs, drawing attention (by display and/or printer) to any abnormal input which might reveal a trouble-spot or the malfunctioning of a sensor.

Additions to the traditional heating and ventilation inputs which have been incorporated in computer systems include the readings of meteorological sensors; operation of lighting, photoperiod covers and thermal screens; irrigation and mist control; control of CO_2 enrichment and NFC solutions. Some of these merit further space here, since they involve more than a time clock, and others are considered in the following sections.

Meteorological inputs   In part, information on the season's weather, stored and then displayed in summarised form on demand, provides the grower with a means to analyse the growth and yield patterns of his crop, for whatever management lessons or insights he may be seeking. In part, the data may be used to generate control signals. In the context of the aerial environment stored data on wind speed and direction can be employed to determine the maximum ventilator opening in given wind conditions and to control the ventilator motors accordingly.

Thermal screens   The simplest mode of operation of these screens consists of drawing them over the crop at dusk and furling them again at dawn, using the solar time clock in computer memory. Unfortunately, a sudden withdrawal of the screens after a cold night may bring down a mass of cold air from the roof zone onto the crop, with adverse effects. Also it is sometimes economically advantageous to keep the crop covered after dawn, so continuing to conserve energy, until the light level tips the balance in favour of photosynthesis. This requires a computer programme linking a radiation sensor (solarimeter or PAR cell) to the screen control system.

CO_2 enrichment The injection of CO_2 must be linked to the ventilator control system to avoid wasteful use of the gas. Again, a programme must be stored in the computer for this purpose.

4.2.3 *Distributed processors*

Even without the considerations of root zone monitoring and control to be discussed in the next section, the computer now has a variety of tasks, which may require data and control links extending over several hundreds of metres in larger greenhouse enterprises. The increasing capabilities of microprocessor-based computers and the need for the utmost systems reliability combine to favour the distributed-processor concept (2.3.2), in which the necessary computing power is placed as close as possible to the scene of operations.

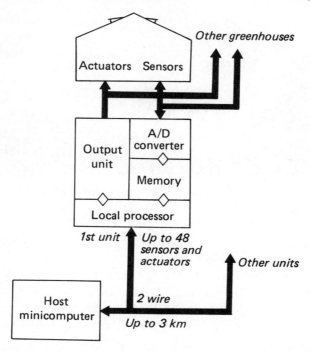

Fig. 4.1 A distributed monitoring and control system for a large greenhouse enterprise (NIAE).

A distributed system for a large greenhouse installation is shown diagrammatically in fig. 4.1. The central, 'host' computer collects and stores all the data gathered from the sensors around the site and supervises a number of remote processors, each of which is responsible for monitoring and control of a discrete sector of the installation (plate 11). The host processor has the VDU and keyboard which enable the manager and his staff to interrogate the system, to change settings and to make other uses of the host, such as running data analysis and business management programmes. Changes of greenhouse conditions are normally sufficiently slow to require only intermittent supervision by the host, therefore it has free time to act in this multi-tasking mode.

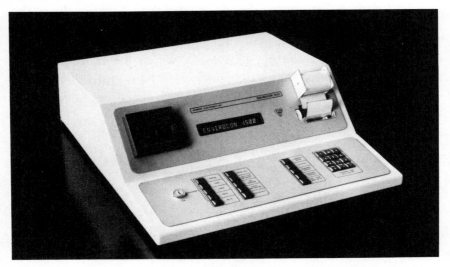

Plate 11 Central monitoring and control station for the greenhouse (Richwest Electronics Ltd.).

The local processors look after their own set of transducers and actuators (maybe up to 48 in all) and are placed where they minimise the lengths of cable involved. Their connection to the host processor is by a two-wire cable carrying serial data. If there is a fault in this link or in the host processor the local processors can operate in the stand-alone mode for as long as necessary, continuing to carry out the programmes in their PROMs or EPROMs. In fact the smaller grower can employ a basic system, incorporating one or more of these local processors, without the need for the host machine.

Diagnostic and alarm systems are essential for such conditions as heating or ventilation failure, and there must be the facility to revert to basic control if the computer system is out of action. The system should have back-up power supplies in case of mains failure.

4.3 Root zone environment

In the commercial greenhouse plants may be growing in soil, peat bags and other solid composts, or in nutrient solutions. Both solid and liquid media call for measurement of temperature, electrical conductivity and pH. Moisture measurement is an additional feature of growing in solids, which need to be irrigated, of course. Apart from irrigation control, soil warming by hot water pipe or electrical cable is sometimes practised and this system requires monitoring and control of soil temperatures.

Measurement of all the quantities just mentioned is outlined in chapters 1 and 3. Control functions can be carried out by the distributed computer systems described in 4.2 or by special units. Irrigation of normal soil can be controlled by computer on the basis of solar energy measurement. Approximate relationships between insolation and water loss from a greenhouse crop have been formulated and the computer can employ one of these to calculate when irrigation water should be applied and for how long. Otherwise water application may be based on soil or compost moisture measurement, using an electrical resistance or capacitance sensor. Special microprocessor-based irrigation control units sequence and time water application and liquid fertiliser injection (plate 12).

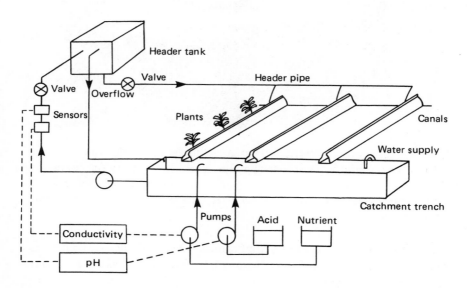

Fig. 4.2 Nutrient film system (NIAE).

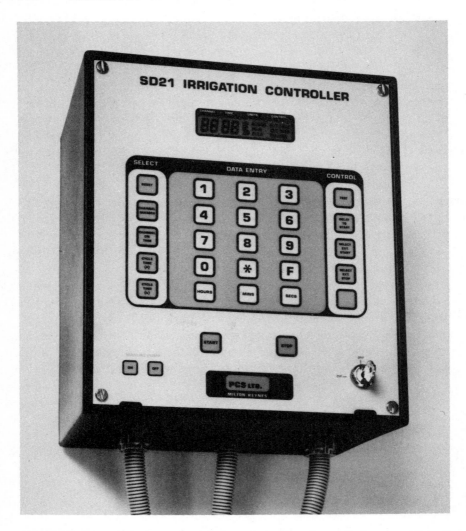

Plate 12 Microprocessor-based irrigation controller for agriculture and horticulture, with capacity for sequencing and timing up to forty operations via a programme keyboard. A security key prevents unauthorised access to the control panel. Test and manual over-ride facilities are included (Peerless Control Systems Ltd.).

An NFC layout is shown diagrammatically in fig. 4.2. The basic monitoring and control functions are needed to measure and maintain the level of nutrients in the solution and to do the same for pH. The controller must monitor for fault conditions, such as no flow, as well. The catchment trench is topped up with water, acid (phosphoric or nitric) and concentrated nutrient solutions, as required by the conductivity and pH sensors, to maintain levels of 2000 to 3000 μS and pH 6.5 to 7 respectively. Liquid temperature is measured, too, and the tank may be warmed to maintain the required plant root temperature, which has an important bearing on plant performance.

More detailed monitoring of NFC solutions is desirable, because any build-up of toxic conditions is liable to be catastrophic to the crop. Nutrient solutions are regularly replaced to reduce this risk at present but new sensors are being developed for improved monitoring of the individual nutrient levels and of the quality of the make-up water itself. Commercial ion-selective electrodes are available for a range of solute ions, including positively charged calcium, potassium and sodium and negatively charged nitrate, chloride and ammonia. These are essentially similar to the glass pH electrode (which responds to the concentration of hydrogen ions in a solution) and they are used with a reference electrode in the same way. They are capable of monitoring individual constituents of nutrient solutions but they need rather more regular care and maintenance than is desirable on a commercial greenhouse site. However, new and more stable sensors, covering a wider range of ions, are emerging from research laboratories to meet the requirements of industry, medicine and agriculture for small, reliable and relatively inexpensive devices. These will interface with computer-based monitoring and control equipment in the same way as the pH electrode, through high-impedance input circuits (see ISFET, 2.2.2).

Checks on the pH, conductivity and ion-selective electrodes can be made with standard solutions supplied or made up to manufacturers' specifications. The more regular these checks, the greater the assurance the grower will have in the accuracy of the equipment.

4.4 Mushroom production

Mushroom culture differs from greenhouse culture in some obvious ways. Most of the operation is carried out in thermally well-insulated chambers and in the dark. Furthermore, the compost in which the fungi grow supplies its own heat through biological and chemical activity. Third, this compost is more variable than those used with greenhouse crops and fourth, yields are less predictable. Nevertheless, from the instrumentation and control standpoint there are many similarities in the type of equipment needed for improved monitoring and control of the phases of mushroom production.

Mushroom growing involves four sequential processes. First, compost is mixed and packed into the trays or other containers which will be used for the rest of the sequence. Next, the trays are stacked in a pasteurising room, to kill pests and unwanted types of fungi. Third, the cooled compost is mixed with mushroom spawn and repacked into the trays before they are put in the 'spawn running' room for 10 to 14 days. Lastly a capping or casing layer is applied and the trays are moved to the growing room. The last three stages require very precise control of environment if the cycle is to be successful and the yield high.

The pasteurisation cycle involves raising the temperature of the compost to about 55°C and holding it at this level for up to a week, or sometimes longer. Self-heating of the compost plays a considerable part in the process and fan ventilation is necessary to maintain temperature control. A drop in heat and ammonia production marks the end of the first phase of the cycle. Steam injection raises the temperature to 60°C at this point and the higher temperature is maintained for several hours before it is allowed to drop to around 55°C again. When release of ammonia is complete the compost is cooled to 25°C before the spawning stage begins. In the spawn running room the temperature is maintained at close to 20°C and the R.H. above 90%. The CO_2 generated is monitored and its level maintained below 6% by ventilation.

The growing cycle has two phases. The pre-cropping phase is associated with a controlled air temperature of 16°C to 18°C and a CO_2 level not lower than 0.1%. Humidity must also be held at around 90% R.H. Uniformity of temperature is particularly important and this involves careful attention to ventilating air distribution. Air-speeds over the beds need to be in the range 0.03 to 0.05 m/s. The cropping phase requires the same air temperature, R.H. somewhere in the region 70% to 95% and a CO_2 level below 0.06%.

It will be evident that efficient mushroom production calls for monitoring of temperature, R.H., air flow, CO_2 and − to a lesser extent − ammonia. Electrical thermometers and hygrometers find application here, as in greenhouses. The same applies to infra-red absorption or conductimetric CO_2 monitors. Ammonia can be measured by a method similar to CO_2 conductimetry. Air is bubbled at a fixed rate into an alkaline solution (sodium hydroxide, pH 12) and the total ammonium ions measured with an ion-selective electrode. However, the cautionary notes in 4.3 about these electrodes still apply. Anemometers of the types described in 1.3.3 can be employed as hand-held survey instruments or installed more permanently.

Collectively, the amount and precision of the monitoring required throughout the last three stages of production provide a further application for the distributed processing system described in 4.2.3, or for basic parts of it to suit the smaller producers.

4.5 Future developments

The high energy costs of heated greenhouses and the risk of continuing dependence on fuel oil will almost certainly drive major producers to locate their greenhouses close to industrial plant with available reject heat, or any other reliable source of cheap power, in regions of favourable winter light intensity. As stated in the introduction to this chapter, there is an active interest in contributions from geothermal, solar and wind power supplies (although the two latter sources are likely to require expensive and extensive energy storage systems to match supply and demand). Heat pumps and systems for heat extraction from the moist greenhouse air may be employed. Greenhouses may fit in with some systems of biogas production and utilisation. Under conditions of high insolation, where cooling is the requirement, refrigeration and moisture conservation equipment will be developed further. All of these systems will provide work for distributed processor control systems.

Experimental 'factory' production of greenhouse crops has shown that high yields and uniform quality can be obtained, using artificial illumination, hydroponics systems for the root zone and production-line methods of growing the crop from seeding to harvest. If these prove economic on a commercial scale they too will involve computer monitoring, control and data management.

In 4.2.2 it was stated that although digital methods have made it simpler to accommodate multi-input systems than the preceding analogue systems allowed, the full potential of the former has yet to be exploited. This statement was based on the fact that biological understanding of the needs of crops for optimum performance throughout their life cycle is very far from complete. The best that has been achieved so far, in the form of recommended 'blueprint' growing regimes, relies on simple rules for the most important factors, such as day and night air temperatures. It is unlikely that these regimes represent the optimum in production control, therefore another significant development in the greenhouse sphere may be the introduction of adaptive control methods.

Adaptive operation entails adjustment of the control system in response to the plants' growth pattern, thereby moving nearer to optimum growing conditions at all times. This assumes the ability of the control system to monitor the plants' performance continuously, or at sufficiently frequent intervals for it to meet the plants' changing needs. For example, maximising CO_2 uptake has been advocated as a means of increasing plant performance. The rate of uptake depends on CO_2 concentration, air temperature and light intensity, inter alia. CO_2 assimilation can be measured by growing test plants in an entirely closed system and monitoring the aerial CO_2 in the system (the 'cuvette' method), but new CO_2 sensors may make it possible to measure assimilation in the plants directly.

If practical plant monitors such as this can be devised, computer control is capable of adjusting one or more of the control settings (heat input, CO_2 injection, etc.) to match the plants' requirements under varying light levels and at different stages of growth.

In several ways, too, methods of statistical analysis and of operational research are increasing the prospect of practical adaptive control to maximise output or − by no means the same thing − the profitability of the production process. The applications of advanced process control techniques to the commercial greenhouse sector (and possibly to mushroom production, too) appear to offer many opportunities for the microprocessor-based monitoring and control systems of the next two decades.

4.6 Further reading

General

Hanan, J.J., Holley, W.D. and Goldsberry, K.L. (1978). *Greenhouse Management. Advanced Series in Agricultural Sciences, 5.* Berlin: Springer-Verlag. New York: Heidelberg.

Mastalerz, J.W. (1977). *The Greenhouse Environment. The Effect of Environmental Factors on Flower Crops.* New York: John Wiley & Sons.

van Koot, Y. and van der Borg, H.H. *eds.* (1975). *Symposium on Protected Cultivation of Flowers and Vegetables.* Acta Horticulturae 51

Section 4.2

Bailey, B.J. (1975). 'Measuring glasshouse heat inputs and limitations of the methods'. *The Grower* **83**, 69-70.

van der Borg, H.H. *ed*. (1980). *Symposium on Computers in Greenhouse Climate Control*. Acta Horticulturae **106**.

von Zabeltitz, C. *ed*. (1977). *Technical and Physical Aspects of Energy Saving in Greenhouses*. Kirchberg, Luxembourg: Commission of the European Communities.

Wadsworth, R.M. *ed*. (1968). *The Measurement of Environmental Factors in Terrestrial Ecology*. Oxford: Blackwell Scientific Publications. Bowman, G.E. 'The measurement of carbon dioxide concentration in the atmosphere' 131-139.

Section 4.3

Bebb, D. (1981). 'Controlling water supply'. *The Grower* **95**, 37-42.

Winsor, G.W., Hurd, R.G. and Price, D. (1979). *Growers' Bulletin No.5. Nutrient Film Technique*. Littlehampton, Sussex: Glasshouse Crops Research Institute.

Section 4.4

Atkins, F.C. (1974). *Mushroom Growing Today* 6th edn. London: Faber & Faber.

Section 4.5

Hand, D.W. and Bowman, G.E. (1969). 'Carbon dioxide assimilation measurement in a controlled environment glasshouse'. *Journal of Agricultural Engineering Research* **14**, 92-99.

5 Crop Handling, Processing and Storage

5.1 Introduction

This chapter covers the range of operations which follow the harvesting of a crop, until it is dispatched from the farm or packing station, unloaded into animal feeding systems or ready for establishment as the succeeding year's crop. The subject falls naturally into two parts, i.e. handling and processing crop material on the move and subjecting the material to various forms of environmental control in store.

The first section deals with the monitoring and control aspects of handling and weighing a bulk crop from the field to the farmstead and subsequently, to the time when it moves on by one or other of the routes given above. This section is largely about weighing, since electronic control of materials handling equipment is not so far advanced in this sphere, in general. Furthermore, materials handling of feedstuffs en route from store to livestock, which could have been included here, is dealt with in later chapters. In the context of the present chapter, weighing must assume particular importance, too, because it generally determines the farmers' and growers' returns on their investments at the end of each season.

The second section is concerned with two other important factors in determining these returns. Consumers are increasingly looking for marketed produce which is clean, free of blemish and of uniform size or weight. Supermarket chains require these high standards of their suppliers. This section therefore covers the applications of electronics to cleaning and grading of crop produce, in which sphere electronics – and the microprocessor in particular – has already found extensive application. In fact, cleaning and grading provide many openings for advanced automatic monitoring and control, making full use of new sensors and the more powerful 'chips' mentioned in chapter 2.

The third section, on crop drying, bridges the two parts of the subject, since some drying is done on the move (albeit slowly) and some in store. Again, the majority of the section is devoted to one subject — grain drying — which is of vital importance to grain producers in climates as liable to summer rain and high humidity as the UK. It also reintroduces the topic of grain moisture determination, already touched on in chapter 3, which also plays a crucial role in determining the farmer's returns. Within the EEC this includes returns on selling into intervention store, of course. Electronics has had a limited role in this sphere for years but it can be expected to become more important, for reasons that will be outlined in the section.

The fourth section is concerned entirely with environment monitoring and control in crop stores. The minimisation of crop deterioration in store is a matter of concern to the farmer and grower, and this can depend on close control of storage conditions. Although electronics equipment is not universally employed in this sphere, it is an obvious one for the application of electronic instrumentation and control, and further developments are inevitable. The same applies to germination and growing rooms, potato chitting houses and any other enclosures used to prepare seeds or seedlings for planting. In addition to coverage of grain, vegetables and fruit stores, a brief reference is made to storage of forage.

Finally, the section on future developments looks at likely trends, mainly in the direction of integrated, computer-based monitoring and control systems.

5.2. Crop handling and weighing

The methods of handling and weighing crops depend so much on the physical characteristics of individual crops that this section is largely based on crop type rather than any electronics classification. The exception is vehicle and trailer weighing, which can be applied to any type of crop being conveyed in bulk.

5.2.1 Weighbridges and weighpads

Weighbridges and portable weighpads are convenient devices for weighing bulk produce, and the former can be sufficiently accurate to meet trading standards. Electronic load cells and their associated circuits, together with digital display and printing of the measured weight, are displacing the old mechanical lever systems of weighing. It is unusual for a farmer to have a permanent weighbridge on his premises, but he can measure his trailer loads with sets of two, three or four unit weighpads, set up on level, hard ground (concrete for preference) at the spacings required by his tractors and trailers. Commercial hydrostatic units are available for this purpose. These have been designed for check weighings on axle loads of road vehicles and they have some disadvantages for agricultural use. In particular, they do not sum the individual readings of the pads, nor do they provide a taring facility. Their dials, which are integral with the pads, can be inconvenient to read, too. By contrast, electronic load transducers allow not only summation of each pad's output, to provide a total weight, but also taring, so that the tractor and trailer can be weighed empty and subsequent weighings indicate the weight of crop material directly. Readings on the separate meter are digital, using light-emitting diodes or liquid crystal displays, and the complete systems run from low voltage batteries.

Electronic weighpad systems are available with load cells rated at 10 tonnes maximum load each. Minimum resolution is usually 10 kg. The accuracy attainable with them is dependent in part on the care with which they are set up and loaded and in part on the design which, ideally, should protect the load cell from all forces in the horizontal plane. Ambient temperature correction may be necessary, too (see 1.3.1). Given pads of good design, properly used, they can achieve accuracies of ± 2% FRO.

As a footnote to this sub-section, it should be noted that the speed and accuracy with which pads can be set up varies widely. Individually, they should not weigh more than about 40 kg, or they become hard to handle. They should also be both wide and long enough to accommodate the largest tyres with which they are going to be used and they should have a low profile (not much higher than 0.1 m).

Checks on weighpads must be made from time to time, of course. This can be done by comparison with the reading of a certificated weighbridge, using precisely the same load.

5.2.2 Grain weighing

One form of grain weigher has been described in 3.2.4. The high rate mass flowmeter developed for attachment to combine discharge augers (fig. 3.4) also has a role in farm materials handling installations for grain, peas, beans and any other particulate material for which it has been calibrated. As stated in 3.2.4, the relative freedom from vibration and the absolute freedom from changes in the attitude of the meter, relative to the vertical axis, all help to improve the accuracy of its readings. In a static grain handling installation, therefore, given a correction made for the moisture content of the grain, an accuracy of ± 2% or better may be attained. A microprocessor-based meter circuit can incorporate all the data needed to correct for moisture content variations, as well as calibrations for different materials.

Another approach to electronic weighing of grain in static installations is provided by the integrating belt weigher, devised as a compact device, capable of working over a wide range of grain flows. Essentially, grain is fed on to a narrow lateral section of a motor-driven, continuous conveyor belt, running at constant speed over two rollers a metre or so apart. The whole assembly — belt, motor, pulley drive and front and rear rollers — is mounted in a frame which is pivoted at the end nearer the grain feed-on point and suspended by two tension springs at the other end. The extension of the springs is proportional to the weight of grain on the belt at any instant and this displacement is measured with an electrical transducer. The transducer's analogue output is converted to a pulse frequency and accumulated in a digital counter, which therefore integrates mass flow. The integrated count is displayed on a digital panel meter.

The commercial version of this weigher, shown in plate 13, has overall dimensions 0.69 m high, 0.79 m wide and 1.63 m long, which enables it to fit into most grain handling installations. The belt itself is 0.61 m wide. The motor drive to it can be seen in the left-hand panel in the illustration. By interchanging pulleys the weigher's range can be altered, to cover the whole span from 7 t/h to 60 t/h. The built-in inlet hopper is directly above the drive motor and the discharge outlet is below the end roller, on the right-hand side in the illustration. Apart from the inlet and outlet the weigher is totally enclosed, to protect it against dust and vermin. The separate control unit (on the left in the illustration) has an electronic, 5-digit decimal readout, giving a resolution of 0.01 t, and an electromechanical counter, reading cumulative weight to 0.1 t. The latter retains its reading in the event of a power failure. The control unit also contains a belt speed monitor to indicate belt slip.

Although the weighers just described are only two of many types designed for grain handling installations, their use of electronics helps to make them versatile and adds the ability to interact with other parts of the handling system.

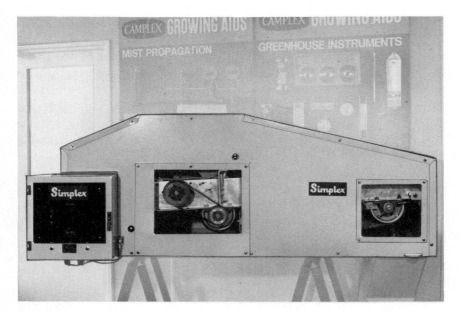

Plate 13 Belt flowmeter for grain-handling installations (Bentall-Simplex Ltd.).

5.2.3 *Vegetable weighing*

The microprocessor has found application in equipment for batch-weighing and packing potatoes, onions, brussels sprouts and other vegetables. Microprocessor control of weighing, filling and sealing bagged produce enables the equipment to attain high throughput rates (e.g. 30 t/h) with close tolerance on weights. The electronics controls the rate of feed of the crop into the weighing containers, reducing it to a trickle as the preset batch weight is approached, in order to avoid overshoot of this weight. The control unit also monitors for malfunctions of the system, such as blockages and the fitting of faulty bags at the discharge point of the machine. The state of all the machine functions monitored by the microprocessor is indicated by digital displays.

5.2.4 *Flow monitoring and control*

The advent of automatic and semi-automatic processing equipment for grain, vegetables and fruit has brought about the need to monitor and regulate the flow of the crop material through the whole handling system. Flow detectors are needed in all sections of the system, to provide the logic signals required at start-up and close-down of the system and to give warning of blockages at any time. In some installations flow monitors also signal the degree of loading of individual sections, as an input to flow rate control systems (see below).

Simple and robust electrical transducers are available to provide flow/no flow indications to the electronic monitoring and control system. Paddle, diaphragm and flap switches all have their place — the first-mentioned in elevator and conveyor lines, the second in feed hoppers as level detectors and the third in all three applications. The sequential control systems of which they are part respond to their signals immediately or after a preset time delay, according to circumstances. For example, stage two of a sequence may not start up until a flow detector in stage one indicates the arrival of crop material; while at the end of the programme stage two may continue for a time after the detector has indicated the end of flow, to clear the material from the system. Similarly, in conditions of irregular flow, the control system can be programmed to ignore a 'no flow' signal from a detector unless it persists for longer than a preset time. All of these operations could be, and were, built into pre-microelectronics systems, with the aid of time-delay circuits. However, the microprocessor provides a means to set up sequences of many different kinds with the same hardware, and to make changes as necessary by reprogramming (possibly by exchanging PROMs or EPROMs).

The loading of a conveyor can be determined optically by a solid-state TV camera or, more simply, by linear photodiode arrays which view a transverse section of the conveyor from above. The output from either of these monitors is a signal which can be used to change the speed of the observed conveyor, or of one downstream, to achieve optimum throughput of the processing stage. Once again, this provides a natural application of microprocessor control.

5.3 Crop cleaning and grading

Mechanical treatments such as sieving and aspiration hold sway in the sphere of crop cleaning as they have done in the past. Equally, the human eye and brain continue to be used to identify individually acceptable or unacceptable objects in a moving layer of produce, using this combination of sensor, computer and memory bank to make often subtle distinctions between the acceptable and the reject for a particular grade quality. The human hand does not necessarily feature as the means to remove accepted or rejected objects, however. This may be done mechanically, on receipt of information keyed in by an operator. Beyond this, automatic systems for cleaning and grading produce have been developed and have made steady progress commercially. Some of these systems are more appropriate to industrial processing plant than to on-farm operations. Nevertheless, they are sufficiently close to the needs of larger farm enterprises to merit some description in this section. The last part of the section deals with a different aspect of crop quality determination, namely protein determination.

5.3.1 *Colour sorting of seeds and vegetables*

Optical equipment for removing discoloured seeds from bulk consignments has been in use for decades. The measurement principle has not changed radically over the years but the electronics has, keeping pace with advances in semiconductor technology. The separators examine each seed in a fast moving stream with an optical viewing system which looks at the seeds from all sides in bright light (quartz-halogen lamps or fluorescent tubes may be used). If there is a sufficient difference between the colour or surface brightness of any seed and a reference standard it is removed by a fast-acting air ejector.

Optical seed separators can work in two modes — monochromatic and bichromatic. The former relies on the existence of an adequate contrast in surface brightness between the seeds of acceptable standard and the remainder. This is the simplest and least costly method of separation. Many seeds do not provide the necessary contrast; there is a range of colour (sometimes wide) within which a seed is acceptable. The defects to be detected may be discolorations overall or in patches. This calls for greater discrimination by the sensing circuits. The bichromatic system provides the means to meet these requirements by simultaneous measurement of the reflected light from seeds in two wavebands (red and green). The equipment is provided with adjustments for colour and shade balance which can be set to achieve optimum separation with any batch of seeds. One machine is capable of working with many types of seed, through the range of adjustment provided. Control over the presentation of seeds to the photometric circuit is achieved by the provision of several types of feed mechanism. Vibratory feeders, chutes, rollers and belts are used, as appropriate to the material to be sorted. The compressed air used for rejecting unwanted seeds is also employed to keep dust from the optical system.

Vegetable produce can be sorted successfully by computer-based colour sorters, too. Conveyor belt feeds are employed. Stones, clods and foreign objects can often be detected and removed from among potatoes, onions and other vegetables by colour difference, while the vegetables themselves are being sorted. Similarly, green and part-green tomatoes can be separated from red ones. With produce in this size range the pneumatic ejector has to be replaced by controlled 'fingers', similar to those used in the X-ray separator for potato harvesters (fig. 3.5). Sorting rates of over 20 t/h are attainable.

There is also evidence that spectral reflectance measurements on potatoes in the red and infra-red wavebands may reveal some diseases of the tubers before they are visible to the eye.

Although the optical density of nearly all agricultural and horti-cultural crops is high it is still possible to use light transmission rather than reflectance as a means for grading produce. For example, grapes, blueberries and other small fruit can be sorted for maturity by measurement of their differential density at specific wavelengths. For black grapes and blueberries the two wavelengths are 740 nm and 800 nm. The fruit must be singled and presented to the optical system sequentially. This system can distinguish both under- and over-ripe fruit from ripe fruit and it uses compressed air to remove fruit in the two former categories.

5.3.2 X-ray sorting

The X-ray sorter described in 3.2.4 and illustrated in fig. 3.5 is equally effective in a potato grading line, if not more so. Its use of trans-mitted radiation, rather than reflection, makes it less likely to err than an optical system when faced with a potato-coloured stone or an earth-covered potato. On the other hand, these X-rays can give no indication of potato quality.

The X-ray system for selecting mature lettuces (3.2.4) can also be applied to the grading line for leaf vegetables generally.

5.3.3 Visual sorting

Electronics aids are available to operators who are examining produce visually on grading lines. These take the form of electronic probes with which they 'mark' each object that they wish to divert into a particular grade channel, by touching it or its image on a TV screen. Figure 5.1 shows a diagram of a system of the latter type, devised for quality grading of potatoes by an operator seated in front of a TV monitor. The monitor displays a picture of the tubers on a roller table, produced by a TV camera (black and white or colour) mounted above the table. The operator touches the TV screen wherever there is a discoloured potato, say, and so activates a trigger signal to the control logic. The pen also picks up the TV scan and by this means relays its position in X and Y co-ordinates to the control logic, through the time relationship between the start of each scan and the pen's detection of it. The position of an identified tuber on the roller table as it advances is memorised by the electronics from that time onwards. This position is retained in memory until the tuber has reached the separator mechanism, at which point the logic signals its Y-address (lateral position on the table) to the bank of deflecting fingers and the appropriate finger deflects it into the reject channel.

5.3.4 Size and weight grading

Automatic size grading by optical measurement is available on some potato graders. These employ linear or matrix arrays of photodiodes on which the tubers are imaged as they pass along a conveyor or down a delivery chute. The passage of each tuber is marked by a change in the level of light falling on the diodes. The number of diodes affected, and the time for which they are affected, provides a measure of the tuber's size.

Automatic individual weighing of larger fruit, such as apples, is readily incorporated in fruit grading lines which present the fruit individually for quality inspection in conveyor cups. The system must be designed to allow each conveyor cup in turn to be completely supported on a load cell as it reaches the weighing point. The output from the load cell is tared to eliminate the weight of the cup itself.

Figure 5.2 outlines a fruit grader system in which quality classifications, judged by eye, and weight classifications are combined, to separate the fruit into multiple quality/weight grades. The operators have a pair of push-buttons with which to allocate one of four grade qualities to each fruit as it reaches them. Weights measured by an L.V.D.T. load cell are classified into seven ranges electronically. The possible combination of twenty-eight different quality/weight grades is reduced to sixteen by a patchboard which consigns as many of the grades as are required to fifteen different intermediate addresses (discharge points), the end of the line acting as the sixteenth address for all non-consigned grades. The discharge address for each fruit is set into two more buttons which travel with each conveyor cup. Each button can be set in one of four height positions by three solenoid actuators (one position being the reset state), thereby providing sixteen possible addresses. A diode matrix and a 'hold' circuit supply the control signals to the solenoid banks for the time needed to programme the push-buttons and then await the next coding signal.

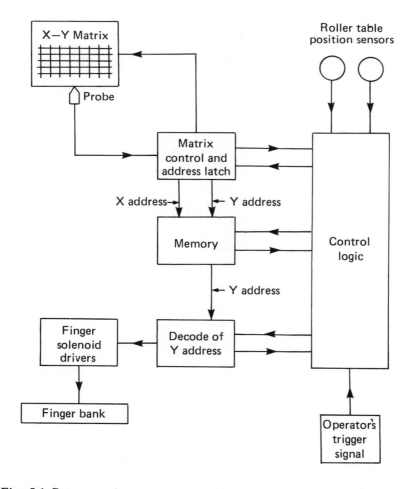

Fig. 5.1 Potato grading using a TV screen and light pen (Loctronic Graders Ltd.).

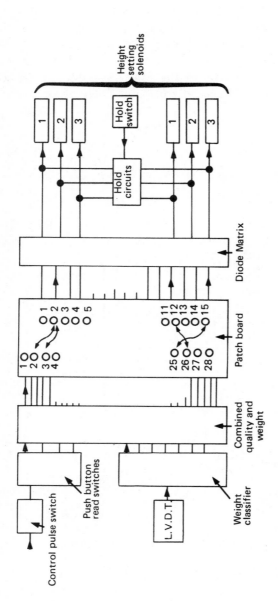

Fig. 5.2 Fruit grading system, combining visual assessment of quality with automatic weight classification (NIAE).

5.3.5 *Complete systems*

Microprocessor-based equipment is employed to monitor and control complete grading lines, starting with control of feed rate and conveyor speed in the preliminary cleaning stages; combining size, weight and grade information and controlling the separator mechanisms. In addition, it can be set up to batch the produce in each grade, by number or weight. At the end of a run it has in memory complete details of the total throughput and the sub-totals in each grade. These data can be aggregated over more than one run, if required, and the manager of the installation is able at any time to call up the information on output that he needs.

5.3.6 *Grain quality*

So far in this section the emphasis has moved from a property of bulk crop material (i.e. weight — or, to be precise — mass) to the properties of single fruit or vegetables which relate to grade quality. On-farm quality determination extends to grain in bulk, however, through moisture determination, already discussed in 3.4.1. Another important factor is the protein content of the grain, which can be determined from the infra-red reflectance of a pre-ground and compacted sample at wavelengths between 1.5 μm and 2.5 μm. Portable electronic equipment working on this principle is used mainly in central storage installations, although it provides information valuable to the farmer and in its solid-state form it could find application on farms.

5.4 Crop drying

It was stated in the introduction to this section that electronics has had a limited role in the crop drying sphere in the past but that it will assume greater importance in the future, particularly for grain drying. There are several reasons for this view. First, although solar energy collectors will be employed increasingly for crop drying in regions of sufficiently intense and regular isolation, hydrocarbon fuels will still be required elsewhere for rapid drying before storage. The need to use these fuels with maximum efficiency is self-evident. Second, unless sunshine and warm, dry air are abundant at harvest time and subsequently, low-temperature drying in store requires selection of the ambient conditions which will be most effective.

Third, the drier itself is often only one part of a complete crop handling system which needs monitoring and control overall for optimum efficiency. Microprocessor-based monitoring and control offers the best route to efficient operation on all three counts. In particular, computer modelling of hot-air drying has advanced to the point at which a computer-based controller — provided with data on the drier and on the input condition of crop and drying air — can calculate the drier control settings which will give optimum drying efficiency.

5.4.1 Rapid drying

Taking high-temperature drying first, the efficiency of electronic control of continuous-flow driers for grain was demonstrated in the 1960s, through application of capacitance and resistance sensing of moisture content and adjustment of throughput rate to achieve an output moisture content as close as possible to 14% m.c.w.b. One form of this closed-loop drier control is shown in fig. 5.3. A capacitance probe with a built-in temperature sensor (thermistor) is inserted in the grain flow (which is vertically downwards) below the heating zone. The probe forms one arm of a bridge circuit, energised at a frequency of about 1 MHz. The output from the bridge, adjusted for grain temperature, is compared with a signal representing 14% m.c.w.b. in the grain to be dried (derived by calibration). The error signal, after amplification, is employed to change the speed of the discharge motor. Sudden changes of the moisture content of the grain at the inlet to the drier can be countered by modifying the control action in response to a signal from an input sensor. For maximum accuracy, the probe has to be placed where grain bulk density variations are at a minimum. Field comparisons of the performance of a drier under manual control and under the control of this equipment in the 1960s showed that the latter considerably reduced over- and under-drying, consequently reducing fuel in one case and risk of grain loss in the other.

The equipment of the 1960s needed separate calibration charts for each type of seed drier but the microprocessor-based controller of the 1980s can refer to stored calibration data in its memory, like some moisture meters (3.4.1).

Its monitoring functions include essential safety features, e.g. warning of burner, fan or discharge motor failure, blockage of crop flow or overheating of the material. It can also store, display and

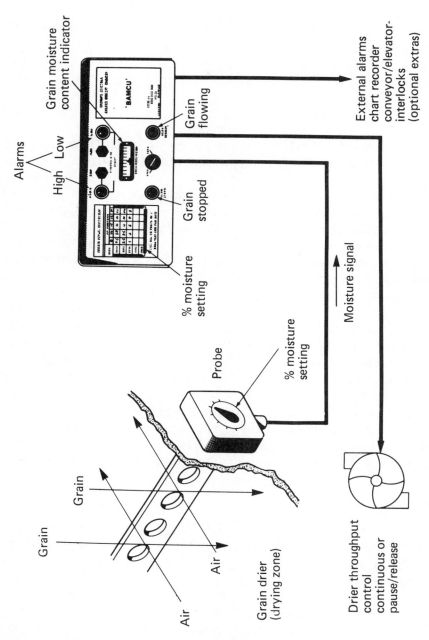

Fig. 5.3 Moisture content controller for continuous flow grain driers, using a capacitative sensor (NIAE).

record data on each drying run, including fuel consumption, input and output moisture contents, and — given a weigher in the system — the mass of material dried. Furthermore, by processing the measured data in accordance with the mathematical models (drier simulations) already referred to it adds greater thermal efficiency to the operation.

5.4.2 *Drying in store*

Effective low-temperature drying of stored material, at or near ambient temperature, relies on ventilation of the material when the temperature and R.H. of the ambient air are such that the air can pick up and exhaust moisture from the crop. Therefore, with the aid of temperature and humidity sensors described in chapter 1 and stored psychometric data, microprocessor-based equipment is able to calculate when to ventilate and when to stop. It can also be programmed with weather simulations, based on meteorological data, to add an anticipatory element to its control function.

A microprocessor-based system has been developed for control of a grain drier which uses warm air extracted from a solar heat collector, together with supplementary heat. Its A/D convertor takes in the outputs from wet and dry bulb air thermometers and from grain moisture sensors in the warm air and the crop, respectively. The converted data provide an input to a control algorithm (sequence of computer steps) which includes a drying simulation programme and tests of the predicted progress of the drying run against damage risks and drying costs. By an iterative procedure the computer identifies the most cost-efficient regime and adjusts the drying air temperature accordingly, through the supplementary heating system.

5.5 Crop storage

Once a crop has been dried to sufficiently low moisture levels for safe keeping and resides in store the emphasis is on stabilising its condition as far as possible through environmental control. This entails heating, cooling and humidification, to combat frost damage, self-heating and moisture loss by the crop, respectively. The products chiefly concerned are grain, potatoes and other vegetables, and fruit, the last-mentioned requiring specially stringent environmental control. The transducers involved have all been discussed in earlier chapters.

5.5.1 *Grain stores*

Bulk stores of grain on-farm are often large and deep. The stored grain is at risk from heating, condensation from moist air and pest infestation — all of which produce a measurable internal temperature rise. It is therefore necessary to check the temperature in a grain store regularly and at several depths. Portable electronic thermo-meters, with digital displays, are widely employed for spot checks, using 1 m to 2 m probes for insertion of the temperature sensor into the grain. However, for larger stores, with contents of high value, continuous monitoring is essential. In these cases an array of rod probes, distributed throughout the crop, will provide the required safeguard, detecting hot spots and switching on ventilating fans. The system shown in fig. 5.4 can employ up to one thousand sensors, linked by one small cable which carries all their signals. The microprocessor-based monitoring and control unit compares each measured temperature with a preset alarm level. On detection of a hot spot it refers to an ambient temperature sensor and if the latter is significantly lower than the former the fans in the vicinity of the hot spot are switched on via a fan control box. This is achieved by allocating each sensor to the appropriate box. A digital display in the monitoring and control unit displays, as required, time, individual sensor temperatures, the preset alarm temperature and the preset scan limit, which ensures that only the sensors in the crop are monitored.

5.5.2 *Storage of potatoes and other vegetables*

Fig. 5.5 shows a bulk potato store diagrammatically. The tempera-ture sensing and fan ventilation system is similar in principle to that just described for grain but cable sensors rather than rods are used. With this crop the aim is to keep its temperature between 5°C and 10°C, depending on the subsequent use of the tubers (the lower temperatures are needed for ware potatoes). The temperature should never go down to their freezing point, i.e. below −1°C, and the humidity should lie in the range 90% to 95% R.H. The safety thermostat in fig. 5.5. overrides the stack sensors if there is a danger of air below −1°C being drawn in from outside the store and passed through the crop. However, cold air at this temperature can be mixed with warmer store air to ensure safe cooling while making best use of the ambient air. Motor-driven shutters (as shown) or proportioning valves are adjusted according to the response of the duct temperature sensor, to achieve the required mixing.

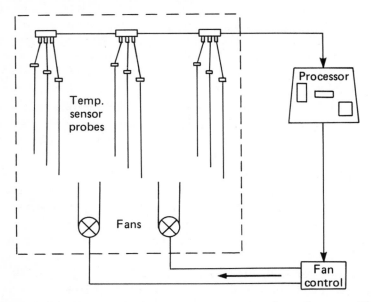

Fig. 5.4 Microprocessor-based temperature monitor for grain stores (Credshire Ltd.).

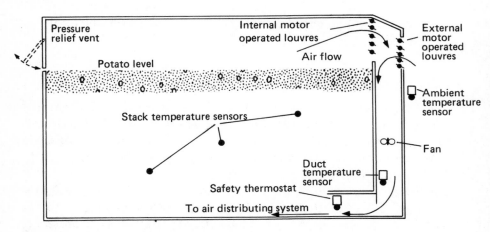

Fig. 5.5 Ambient air mixing/recirculation unit for potato stores (Electricity Council).

In addition to these facilities the complete potato storage installation may employ supplementary heating, particularly for seed potatoes before planting, while refrigeration facilities are needed for long-term storage. In all, this calls for a microprocessor-based system to provide comprehensive temperature and humidity monitoring, display and/or recording, combined with control of ventilation, heating and cooling systems.

Brassicas, carrots and dry bulb onions require storage at or near 0°C. The R.H. should be 70% to 80% for the onions and 95% to 98% for the rest of these crops. Some are stored in buildings with ice-banks, generated by refrigeration units. In these buildings the stored produce is cooled and kept moist at the same time by air which has passed through water chilled by the ice. Tomatoes require a temperature of about 8°C and an R.H. of 85%. For these crops, too, temperature and humidity transducers described in 1.3.4 and 1.3.5 find application, coupled with microprocessor-based monitoring and control to maintain the required environmental regimes.

5.5.3 *Fruit stores*

Refrigerated fruit storage introduces another set of requirements, especially stringent in the case of gas-tight, controlled atmosphere stores. In the CA store restricted ventilation maintains carbon dioxide concentration at well above normal atmospheric levels (reaching 1% and above) and allows oxygen depletion to 3% or below. Nitrogen increases correspondingly. CO_2 is regulated by scrubbing the recirculated air in lime, by flushing with nitrogen or by burning propane in a controlled air supply to generate a gas rich in CO_2 and low in O_2, which is fed into the store after cooling. Control of the ethylene generated by the crop and by any mould growth present is important, as is water loss by the crop. Temperatures must be low but not freezing. These are normally maintained to within ± 0.5°C about a recommended value below 5°C. Once again, there is an obvious application for microprocessor-based monitoring, control and data processing systems.

The instrumentation required for these stores naturally includes temperature and humidity sensors, and instruments for CO_2 and O_2 measurement. The temperature sensors must be accurately calibrated within ± 0.5°C, of course, and carefully located in relation to the cooling system if the temperature throughout the whole store is to be maintained within the required limits. Electronic humidity sensors are satisfactory in this application if protected by a filter. Carbon dioxide and oxygen are measured by one of the appropriate standard industrial methods, namely infra-red absorption or thermal conductivity for CO_2, and paramagnetic measurement for O_2. Ethylene occurs at a very low level (about 0.1 v.p.m.) and there is no comparable instrument for on-line measurement as yet.

5.5.4 *Forage stores*

The important forage crop cannot be entirely ignored here, although the role of electronics is very limited in this context. Temperature and humidity measurements have a place in ventilated drying and storage of hay, which provides an application for simple electronic monitoring and control systems. There is no comparable application to ensiled crops, however.

5.6 Future developments

Comprehensive monitoring, control and data processing systems for crop handling, processing and dispatch are likely to be a standard feature on larger farms. Automatic inspection and grading of produce in the optical and other electromagnetic wavebands will develop further and may become more common on the farm or in the packing station.

Many other applications for microelectronics systems exist in the fields of crop processing for human and animal feed, and in new fields such as micropropagation and bioengineering, but it seems probable that these opportunities lie beyond the 'farm gate'.

5.7 Further reading

Section 5.2
(1979) *Proceedings of Conference on Weighing and Force Measurement. 'Weightech 79'.* London: Institute of Measurement and Control.

Section 5.3
Gaffney, J.J. *ed.* (1976) *ASAE Publication 1-76 Quality detection in foods.* St Joseph, Michigan: American Society of Agricultural Engineers.
Gale, G.E. and Holt, J.B. (1977) *DN 750 The NIAE Fruit Grading Rig, 1976.* Silsoe, Beds.: National Institute of Agricultural Engineering.

McClure, W.F., Rohrbach, R.P. (1978) 'Asynchronous sensing for sorting small fruit'. *Agricultural Engineering* **59**, 13-14.

Porteus, R.L., Muir, A.Y. and Wastie, R.L. (1981) 'The identification of diseases and defects in potato tubers from measurements of optical spectral reflectance'. *Journal of Agricultural Engineering Research* **26**, 151-160.

Section 5.4

Aguilar, C.S. and Boyce, D.S. (1966) 'Temperature ratios for measuring efficiency and for the control of driers'. *Journal of Agricultural Engineering Research* **11**, 16-23.

Gough, M.C. (1976) 'Remote measurement of moisture content in bulk grain using an air extraction method'. *Journal of Agricultural Engineering Research* **21**, 217-219.

Harrell, R.C., Allison, J.M. and McLendon, B.D. (1979) *ASAE Paper No. 79-5524 Microprocessor-based Control System for Solar Assisted Grain Drying*. St Joseph, Michigan: American Society of Agricultural Engineers.

Hawkins, J.C., Messer, H.J.M. and Lindsay, R.T. (1978) 'Cooling of vegetables with positive ventilation and an ice bank cooler'. *ARC Research Review* **4(2)**, 34-37.

Holtman, J.B. and Zachariah, G.L. (1969) 'Computer controls for grain driers'. *Transactions of the American Society of Agricultural Engineers* **12**, 433-437.

Lindsay, R.T. and Neale, M.A. (1975) 'Cooling produce in large pallet-based boxes'. *Journal of Agricultural Engineering Research* **20**, 235-243.

Matthews, J. (1963) 'Automatic moisture content control for grain driers'. *Journal of Agricultural Research* **8**, 207-220.

Matthews, J. (1964) 'Performance of an automatic moisture control unit fitted to a farm drier'. *Journal of Agricultural Engineering Research* **9**, 180-187.

Nellist, M.E. (1976) 'Drying of hay and grain'. *The Agricultural Engineer* **31(3)**, 55-60.

Section 5.5

Ministry of Agriculture, Fisheries and Food (1979) *Reference Book 324 Refrigerated Storage of Fruit and Vegetables*. London: Her Majesty's Stationery Office.

6 Pigs, Sheep and Poultry

6.1 Introduction

The livestock industry is extremely varied, ranging from the intensive to the extensive, from the high capital input to the low. The requirements for livestock engineering in general and electronics in particular vary correspondingly. There is a clear progression from sheep production, which makes few calls on the livestock engineer, through beef cattle, pigs and poultry, to the dairy industry, which is already a major user of electronics equipment. For convenience, therefore, the livestock sector is covered in two chapters, the second of which deals with the advanced dairy farm and those aspects of electronics which also have application to beef cattle. The present chapter deals with the smaller animals and the birds which constitute the other main classes of farm livestock. Mostly it is concerned with larger intensive farms, with minimum labour, where safe, low-voltage electronic systems help with the chores, maintain vigilance over the operation and provide timely management information.

The applications of electronics to the pig, sheep and poultry industries fall into six classes, which are dealt with in turn, before a forecast of future developments. The first of these is monitoring and control of environment in pig and poultry housing to provide the conditions for maximum production. These two types of stock, unlike sheep, are commonly raised intensively in enclosed buildings, with control of ventilation, heating and, in many cases, light throughout the life of the stock. Ventilation serves not only to remove pollutants and moisture from the animals and their waste but also to regulate air temperature by removing some of the heat generated by them and their surroundings. The air movement is created by fans and the desired uniformity of temperature must be maintained in the presence of internal obstructions, such as low dividing walls in pig houses and tiers of cages in poultry laying houses. Additional, controlled heating is needed for young stock, while special and precise environmental conditions are required in egg hatcheries.

Feed costs amount to between 75% and 80% of the total production costs in pig and poultry enterprises, therefore precise control of feed rations is of the utmost importance. Although dry feeding of pigs is often carried out by hand, automatic systems for conveying both dry and wet feeds are gaining ground. Wet feeding can give better conversion rates (i.e. feed to meat) and is likely to become more important in consequence, although it costs more in terms of fixed equipment. Electronic control systems must therefore cope with the delivery of feed in either form. In intensive poultry houses feed is distributed in dry, granular form by conveyor, whether the birds are layers or 'broilers'. Water consumption by pigs and poultry must also be measured if total input is to be monitored.

The liveweight gain of meat animals is also an important economic factor. Regular weighing of animals and birds provides the information on which regulation of feeding can be based, to attain the desired market weights at planned times, for maximum financial returns. Manual weighing is a time-consuming operation, beset by problems if the livestock become alarmed and resist being weighed. In particular, this may affect their conversion rate itself. Electronic weighing makes it possible to minimise these upsets and to improve on the accuracy of manual weight readings. In the case of pigs and sheep, ultrasonic equipment provides another indicator of performance, namely the fat/lean ratio, by in vivo testing.

Ultrasonic measurements play another role in the sheep sector, as a means of pregnancy detection. A short section is devoted to this subject.

Egg production is largely covered by the section on environment and feeding. Egg collection is the preserve of mechanical equipment, designed to minimise damage during conveying and handling. Electronics only finds application at the grading stage, but here it provides a useful aid to efficiency.

Lastly, full utilisation of livestock excrement is being sought in several ways. Animal manure or 'slurry' contains valuable nutrients which can be recycled to the soil, or into animal feedstuffs after careful processing. It can also be put into digestion plant, where microbial action converts some of it to methane gas. The efficiency of this biogas production is highly dependent on the quality of the process control. In consequence, this is another active area for microprocessor monitoring and control.

6.2 Environmental monitoring and control

In the preceding section the relationship between environment and production was mentioned. In particular, there is a well-defined range of ambient temperatures within which livestock perform best, and the economic advantages of establishing these conditions are substantial. Therefore, the best livestock buildings are designed for this purpose, which requires them to be thermally well insulated and leakproof, with controlled ventilation and heating. A selection of the temperature and other environmental requirements of pigs and poultry will indicate the levels of control needed.

Animal scientists have defined the temperatures which bound the pig's 'comfort zone' as the lower critical temperature (LCT) and the upper critical temperature, which is around 30°C. Below the former the animal uses much of its food energy to keep warm, while above the latter it again uses up energy trying to keep cool. The LCT depends on several factors, including the animal's size, feed ration and proximity to other animals, and to its insulation from draughts and cold floors. Published figures on the LCT for a 50 kg pig (heat output about 200 W) give 18°C as its basic value when the animal is given an ad lib feed ration. This value rises if the pig is smaller, if the ration is restricted to provide a lean carcase, if the stocking density is low or if the house is draughty; it decreases if the pig lies on straw bedding and its pen is roofed over. These factors can alter the effective LCT by as much as ± 5°C. Clearly in an unheated house in cold weather it pays to have the right stocking density, thermal insulation and air movement if each animal is to remain within its comfort zone. Newly-born piglets are in need of temperatures of 27°C to 30°C and this requires additional heating in 'creeps', to which they are attracted by light from the heater itself or from a supplementary lamp. The R.H. in pig houses can normally vary fairly widely without serious consequences, but extreme values — say outside the range 50% to 80% — are usually avoided.

The environmental needs of housed poultry are also dependent on age, although there are constant features in the regimes recommended to farmers. Ventilation must be uniform and gentle; background air temperatures should be uniform, even when local heating is employed for chicks at the brooder stage. Relative humidities between 60% and 75% provide the best growth conditions and help to reduce respiratory disease. Temperature-controlled radiant heaters for brooder heating should give local temperatures of about 46°C for day-old chicks, against a background of 25°C to 30°C. Temperatures should be reduced for about three weeks until the post-brooding temperature range is reached. For broilers this is 18°C to 21°C. Intensively kept laying birds require somewhat higher temperatures, about 21°C at least, to maintain optimum egg production. Temperatures above 24°C reduce the numbers of eggs laid and their weight and quality.

Intensive poultry are housed in light-tight buildings and lighting programmes are shaped to stimulate the hens' laying patterns by simulating spring and autumn conditions. These programmes extend or contract the number of continuous light hours in each day up to a maximum of 16 to 18 hours, making changes at weekly intervals or longer, over a 49-week period. Broilers are given more light but have at least one hour of darkness daily to accustom them to this condition as a safety measure. Without this they might panic, crowd and suffocate in the event of a light failure. Sometimes cyclic patterns of four hours' duration are employed, in the interests of improved feed conversion.

Electronic monitoring instruments of several familiar kinds are recommended for use by the manager of the larger pig and poultry enterprise, which can justify the expenditure through the increased profitability of properly maintained environmental conditions. The speeds of ventilating fans can be measured with digital tachometers; airflow by vane-type anemometers or, at low flows, by the cooling effect on hot-wire or thermistor anemometers; temperature by thermistors or electrical resistance thermometers. However, electronics has a greater part to play in on-line control of fan ventilation of livestock buildings, together with control of supplementary heating and lighting, given proper care of the sensors in an aggressive environment. One form of automatic ventilation with wide application to pig and poultry housing is described below.

6.2.1 *Automatic fan and ventilator control*

One of the practical difficulties with fan ventilation of animal houses arises from the wide range of airflows needed to sustain the required internal air temperature, irrespective of outside ambient conditions. If variable-speed fans are employed to cover this range there may be difficulties of airflow control at low fan speeds, particularly in windy conditions. One solution to this problem, applicable in larger houses, is the installation of a series of fans which are operated only in the fully 'on' or fully 'off' condition. Regulation of airflow is then a matter of which fans are switched on at any time. Another problem, perhaps more serious, is the way in which the circulating air movement within a house may alter as external weather conditions change. A stable airflow pattern within the house is essential to accurate control of the livestock environment.

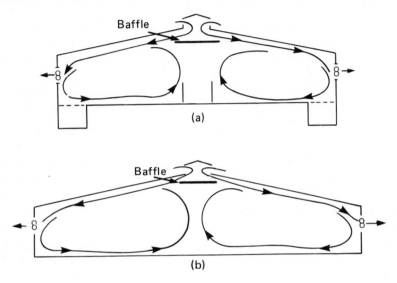

Fig. 6.1 Ventilation system for pig and poultry houses with coupled fan and ventilation control: airflow patterns (NIAE).

The automatically-controlled ventilation system mentioned above employs the stepped fan method and overcomes the air pattern problem through the design of the building's air inlet. Figure 6.1 shows the outlines of two buildings with extractor fans in the side walls and an air inlet in the ridge. This common design can support the stable air patterns shown if the incoming air travels unobstructed towards the fans, setting up the rotational circulation which brings cooler air to the stock at floor level. Therefore the ceiling of this type of house must have a smooth inner surface, clear of purlins and other obtrusions which could deflect the airstream from the desired path. The second requirement (which holds with temperatures down to $-2°C$) is that the inlet airspeed is maintained at a constant level, close to 5 m/s, irrespective of the ventilation rate. This can only be achieved by altering the inlet area whenever the ventilation rate changes and that requires automatic adjustment of the baffles which run the length of the ridge, coupled with the stepped fan control already mentioned.

Fan control is based on the relationship between outside air temperature and the ventilation rate required to maintain stock in their comfort zones as far as possible, calculated from stocking density, animal weight and the known thermal characteristics of the building. This relationship is generally of the form shown in fig. 6.2, from which example it will be seen that a 10:1 range of ventilation rate can cover the temperate range of outside weather conditions. The numbers 1 to 6 indicate the possible levels at which fans are switched in, starting with, say, two constantly running fans (stage 1) and ending with all fans running at stage 6. The greater the number of fans employed the more closely it is possible to match the above relationship. Also it makes it possible to distribute more uniformly the fans which are brought into action at each stage, so helping to prevent the build-up of contaminants in any part of the house. (Note: The number of fans running does not affect the uniformity of distribution of the inlet air along the ridge.)

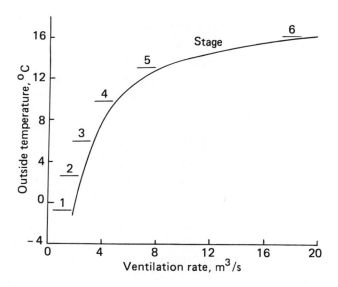

Fig. 6.2 Typical relationship between outside air temperature and internal ventilation rate for an intensive livestock building, indicating the switching levels for the ventilating fans (NIAE).

The adjustable inlet baffles must be of rigid construction and able to withstand warping under extremes of air humidity and condensation, since they are required to establish precise inlet gaps down to small openings. They are raised and lowered on suspension cables which ensure that they remain in the horizontal plane at all settings.

The control system takes inputs from distributed thermistors, which provide an average of the internal air temperature, and operates the fans in the preset sequence, at the same time actuating a reversible electric motor which raises and lowers the baffles, via the cables. A feedback of the position of the baffles can be derived from an angular displacement potentiometer, operated by the motor shaft or the cable.

An important feature in any automatic system upon which livestock depend is the fail safe procedure. In this case, power failure or excessive internal air temperatures must be countered. This is achieved with electromagnets which are de-energised under either of these emergency conditions. One releases a loop in the winch cable which operates the baffles, causing them to drop to the fully open position. Another releases a weight which opens the backdraught shutters fitted over the fan outlets. These would normally shut on loss of power to the fans, of course.

This system has contained the temperature lift in a piggery to 3°C above outside ambient in warmer weather and allowed a 25°C lift (produced by heat from the livestock) under very cold conditions. Night-time temperature fluctuations internally can be as low as ± 0.5°C, rising to about three times that amount during the daytime activity of stock and stockmen. Referring to fig. 6.1, the air circulation in house (a) produces slightly cooler conditions at the dunging passage and warmer ones in the lying area. This clearly coincides with the pigs' preferences and they consistently use the two areas for their intended purposes. The result is cleaner, healthier pigs. If houses have a central dunging passage, air must be introduced horizontally at the eaves and extracted at the ridge, thereby reversing the flow. The turkey house layout in (b) produces almost uniform conditions over the whole floor area, thereby avoiding damp litter near the walls.

Given control of the temperature at livestock level, via the correct air flow pattern, the next logical step is to seek ways of measuring and controlling the air speed over the stock, too, since the optimum temperature for production is dependent in part on this quantity, as already indicated. However, air speeds are low round the stock and they must be measured by sensitive anemometers. Hot wire anemometers have the required sensitivity but the elements are fragile and their response is affected by dust and other deposits. More robust thermistor bead anemometers can provide acceptable sensitivity, with omnidirectional response, but they must still be guarded from mechanical damage while avoiding protective structures which seriously affect the air flow.

In the case of pigs a relationship between air temperature, air speed, pig weight, feed intake, stocking density and LCT has been established by theory and experiment. Weight and feed measurement are discussed in 6.3 and 6.4. If air speed can be measured satisfactorily microprocessor control will combine all these elements, to provide optimal environmental regimes.

In the poultry sector another necessary advance is the modification of the air flow and temperature control system just described to suit layer houses with tiered cages. If this can be done, comparable environmental control systems will be possible.

Summarising, animal science has provided the information and models upon which to base control of air movement and temperature in piggeries and poultry houses. R.H. may have to be kept in bounds by ventilation and heating, for health reasons. (Note: The environment is hazardous to absorption type sensors.) Ventilation also reduces the risk of toxic concentrations of pollutants, among which ammonia and carbon dioxide are important. The relationships between temperature, air movement and other factors influencing productivity, such as body weight, can be built into the environmental control system, with the aid of the microprocessor. The system must always fail safe, in ways such as those already described, and should operate alarms in so doing. Warnings of fire and unauthorised entry into the buildings are readily incorporated. All of this implies the presence of an auxiliary power supply, for the monitoring equipment at least, to provide automatic back-up in the event of mains failure.

This summary would be incomplete, however, if it did not also stress the peculiar risks attending the introduction of measurement transducers within reach of livestock. Defence of the space around each transducer and protection of the cables leading from it are imperative. Considerable ingenuity is needed to protect the equipment without destroying the condition that it is intended to monitor (vide thermistor anemometers). Plate 14 shows an instrument cage designed to survive the attentions of a broiler flock.

Plate 14 Protective cage for recording equipment in a broiler house (NIAE).

6.2.2 *Hatcheries*

Egg hatching in incubators is a mass-production process which calls for precise control of temperatures. Hatchability can change by as much as 10%/°C. It also requires a moderate degree of control over R.H., O_2 and CO_2 levels.

Fumigated eggs arriving from a farm may have to remain in store for some days before it is their turn to move into the incubator itself. Storage must be at a temperature lower than 20°C, to suppress embryo development, and is often between 13°C and 16°C. The humidity is maintained at 75° R.H. In the incubator the eggs remain for a further period at a temperature close to 38°C and within a specified R.H. range between 50% and 60%, the precise level depending on the type of egg and the incubator's design. The R.H. has to be adjusted to limit the water loss from the eggs to between 10% and 12% over the incubation period of about 18 days. This loss can be checked by regular weighings.

The air flow within the incubator must be sufficient to maintain the oxygen level above 15% (it is about 21% in normal air). The CO_2 must also be restricted to a value between 0.1% and 0.4%. Embryo damage will occur at abnormally low or high oxygen concentrations, while the carbon dioxide concentration affects the pH of the egg albumen, which should be between pH 7.2 and pH 7.4.

Immediately before hatch the R.H. is reduced to allow freer gaseous exchange to the unhatched chick — now changing over to lung breathing. It is increased again to about 70% during hatching and then reduced to allow the new-born chicks to dry off.

As in crop stores, electrical resistance thermometers and hygrometers can be used, together with CO_2 meters of the infra-red or thermal conductivity type. The electrical conductivity method described in 4.2.1 is applicable, too. Paramagnetic oxygen meters can be used to measure oxygen concentration if required. Automatic weighing of the incubating eggs will provide the basis for regulation of R.H. Then the whole process — including the repeated turning of the eggs during incubation, as well as temperature and ventilation control — can be brought under integrated microprocessor control.

6.3 Feed monitoring and control

In the main, automatic rationing of both pigs and poultry under intensive conditions is based on volumetric dispensing of dry feed by pipeline or open conveyors, or by travelling hoppers. Measurement of this economically vital input is often based on the time of distribution, which assumes that the flow rate of the materials is constant. This is rarely so with granular feed mixtures. As a first step to improved management control, some farmers have introduced load cells in the support structures of bulk bins. Properly installed, these can provide a moderately accurate measure of the daily intake of the stock. Water meters of the turbine or positive displacement type will provide a measure of daily water consumption, in electrical form. These two inputs can be collated in any central monitoring and control system embracing the environment and/or feed conversion (6.4). The weight of the feed itself is not the important factor but on-line methods of analysing feed quality are not yet available. Therefore, for more precise control of rations, weight must be related to major constituents of the feed — such as protein and ME (metabolisable energy) — using data from suppliers or independent analytical laboratories.

Gravimetric pipeline dispensers of dry and wet feed to pigs have been in use on a small scale for several years, and provide an accuracy of about ± 2%. These are electromechanical or pneumatic in operation, rather than electronic. The material is fed along a pipeline fitted with feed dispensers at intervals. A mechanical weight trip in each dispenser routes the flow on through the pipeline to the next dispenser once the preset weight has been attained, with the result that each fills in sequence. When all are filled electrically-operated solenoids or pneumatic actuators release their contents into the feed troughs below them. Because hungry pigs soon learn to detect the sounds of food on its way and can become very restive at any delay, the hoppers may be filled with the next feed while the animals are still eating the food just dispensed. Electronics can only improve on this system at some cost but the use of a simple electrical displacement transducer in the weighing mechanism of the dispensers (see 6.4) provides the input required to programme each dispenser individually from a central point. In a large pipeline installation this can save time and error, as well as fitting in with central data logging and processing.

A simple and rapid method of mixing water and meal to make a liquid feed is provided by the vortex mixer, in which the feedstuff, falling down a tube, is caught up in a vortex produced by water introduced tangentially into the base of the tube. Control of the water/meal mix is important on two counts: first, uptake by the animals depends on the constitution of the feed and, second, the efficiency of pumping along pipelines is affected. The required control can be obtained by measuring the mass flow of meal into the tube and adjusting the water flow in proportion. A miniature version of the electronic proportioning weigher to be described in 7.3.1 can be employed for this purpose.

Poultry feeding has not developed in the same way. Wet feeding is rarely used and the main concern with dry feeding is to provide uniform distribution and minimum waste, rather than precise weighing of feed to each group of birds. It seems likely, therefore, that electronics will only find application at the bulk storage end of the feeding system. However, should something akin to the gravimetric pig feeding system be called for, the necessary electronics measurement and control techniques have been established.

6.4 Weighing

6.4.1 *Pigs*

Until the arrival of microelectronics-based instrumentation the weight of a pig was almost universally measured with a spring balance, loaded by a weighcrate containing the animal and mounted on an external frame. The accuracy of this system, still widely used, relies on an operator's ability to read the position of a mechanical pointer on a circular or linear scale but determination of this position can be very difficult if the animal is restive or agitated, even when some form of damper is applied to the balance movement. Furthermore, weighing large numbers of pigs in this way can be time-consuming and may need two people to do it — one to drive the animals into the weighcrush in turn and the other to take the weight readings as accurately as possible. Pig weighing was therefore carried out too rarely in the past, thereby depriving the farmer of the regular information on liveweights that provides a check on feed conversion and the approach of each animal to its market weight. Electronic systems have been developed to overcome these problems by providing automatic measurement of weight, which can be coupled with pneumatic control of the weigh crate itself to reduce (but not abolish) the labour input. The essential features of a one-man, semi-automatic pig weighing system are shown in fig. 6.3. The weigh crate is the third element in a four-element layout. The pigs to be weighed are driven into the first element, a holding area, before the weighing starts. The operator then separates one animal and steers it into a holding pen, closing the gate of the pen behind it. While the animal is in this pen the operator can read its identity (normally a number on an ear tag) if required and record this number. Then he opens the guillotine inlet gate to the weigh crate itself and the pig enters, voluntarily or with encouragement, if needed. The inlet gate is closed behind it and it is weighed electronically. While the weighing is in progress another pig is brought into the holding pen. When the first pig's weight has been noted or recorded it is released through the exit gate of the weigh crate into the fourth element of the layout (the post-weighing area) and the second pig takes its place. This sequence is continued until the batch of animals has been weighed and is held in the post-weighing area.

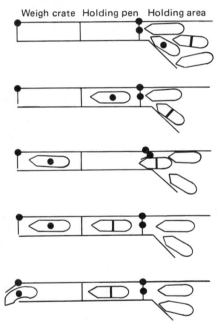

Fig. 6.3 The sequence of operations in one-man weighing of pigs (NIAE).

The whole weighing operation can be performed by one man because the opening and closing of the inlet and outlet gates of the weigh crate is controlled by pneumatic rams. The operator stands at the entrance to the holding pen (plate 15), feeding the pigs into the pen singly through its folding gate. He operates the ram on the guillotine inlet gate via a pneumatic push switch on the frame of the gate, and also operates the ram on the exit gate by pulling on a rope, equally close at hand, which actuates another switch. The floor of the weigh crate is pivoted at the inlet end and rests on a stirrup at the exit end. When the exit gate is opened and the weighed pig leaves the stirrup operates a pneumatic valve, which immediately closes the exit gate again and then opens the inlet gate. The push switch on the guillotine frame is used to reclose the latter gate once the next pig is in the weigh crate. The pneumatic system is powered by a conventional farm compressor.

Plate 15 One-man, semi-automatic pig weigher, with holding pen (NIAE).

Several refinements to this basic arrangement are necessary for smooth and rapid operation of the system. A small gap is left at the foot of the guillotine gate and the waiting pig tends to nose into it, ready to follow the animal already in the weigh crate. A metal barrier prevents it from interfering with the weighing. The sides of the crate itself are blanked off and are closer together at the inlet, so encouraging the pig towards the exit gate and preventing it from turning round. Third, a spring-loaded auxiliary exit gate is mounted on the fixed outer frame of the weigher. This is pushed open by the pneumatically controlled main exit gate on the weigh crate itself and the spring returns it to the frame when the main gate closes again. Its object is to prevent animals already weighed from interfering with the subsequent weighings.

The mechanical details of the system have been described at some length in order to make two general points. First, the problem of singling individuals from a group in order to make a measurement of some kind on each is often difficult enough with crops such as potatoes and apples. When livestock are the subjects it is even more difficult to do this reliably and without danger to the animals. The presence of the human operator allows the singling to be done manually and provides a safeguard if an animal should get into difficulties. Second, as noted in the preceding two sections of this chapter, animal behaviour has a strong influence on the design and operation of monitoring and control equipment for livestock operations. In practice, learning to adapt new designs to the behaviour of the animals concerned frequently takes up more development time than anything else.

In the present instance the one-man weighing system has been proved over a period of years. With it the man can weigh about 100 pigs/h, by handling them in batches of 10 to 15. With a helper to fetch and return these batches a rate of 200/h can be attained.

The electronic weighing system is necessary to sustain the throughput rate and to provide the necessary accuracy. The input to the weight measuring circuit is a linear displacement transducer, fitted in parallel with the spring balance movement. The normal damper on the balance (which is unidirectional) is replaced by units giving light, bidirectional damping. The signal from the transducer is then averaged electronically for two to three seconds before being displayed on a digital meter. By proper scaling of the output signal the meter reads directly in kilograms. Averaging of the signal virtually eliminates the effects of pig movement even when the animal is lively. The steady display which appears after the integrating period makes reading of each animal's weight a simple matter and it has been shown that the displayed weight is normally within 1 kg of the actual weight, determined from more accurate weigh platforms. An accuracy of ± 1% in weighing a 100 kg pig is ample, since its weight can easily change by 1 kg in a short time, through defaecation and urination.

Another feature of the electronics is a comparator circuit which produces an output whenever a pig's weight is above a preset value. The output can be employed to actuate a spray marker mounted above the weigh crate, to identify the pig while it is still in the crate. This is an aid to selection of pigs ready for market. A second output from the meter actuates a printer, if hard copy of the pig weights is needed for farm records.

One cautionary note is needed before leaving this topic. The weight of each pig varies appreciably throughout the day (5% or more), because of its daily ingestion and excretion functions in particular. The accuracy attainable with electronic weighing is therefore only fully exploited if all weighings are taken at comparable times of day or if weighings are corrected to a standard time, based on experiment.

6.4.2 *Poultry*

The need for regular checks on body weight is equally important in the poultry industry. The broiler producer needs to know the growth rate of each crop, partly as a means to check feed conversion and partly to forecast the average bird weight and its statistical spread near marketing dates. The owner or manager of breeding flocks needs to adjust feed input to produce the bird weight which maximises fertile egg production. Singling and weighing birds by hand is time-consuming and causes stress in the birds. Equally, it is not possible to weigh all the birds in a large flock. Therefore, a means of taking sample weights automatically and harmlessly has been devised. This is based on the provision of strain-gauged perches at intervals throughout the poultry house. An inexpensive, low-range load cell is needed for this purpose and the vacuum deposited, cantilever strain-gauge transducer shown in plate 1 (1.3.1) is employed. This has a load capacity of 50 N (equivalent to 5 kg weight approximately).

The design of a practical perch weighing assembly is shown in plate 16. Inevitably, it is the outcome of a long period of development, to determine a design and location which will attract a representative sample of birds (male and/or female) to perch on it, whatever the stage of development of the broiler or breeding flock. Three main factors influenced the design. First, a low-level, flat, rectangular perch is needed, to allow less mobile birds to mount it. Second, it must not swing appreciably or the birds will not settle and the force on the transducer will fluctuate wildly. Third, having settled, the birds must not stay settled for long. This implies that some movement of the perch must be possible, since it has been found that movement by other birds in the vicinity of the perch can be enough to unsettle a perching bird. The stirrup-shaped perch is suspended on a long arm, supported by the cantilever load transducer at the top. It is fixed by a bracket attached to one of the stanchions in the house, with the rectangular platform 30 mm to 60 mm above the floor. It will not rotate but has some lateral flexibility, to limit perching times. It also has a safety spring in the suspension to prevent accidental overload of the transducer.

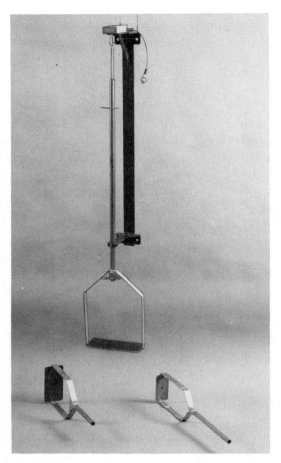

Plate 16 Strain-gauged perch for poultry weighing (NIAE).

The output from each perch goes to a multiplexer, A/D converter and thence into a microcomputer with a digital meter or VDU. The computer first stores a tare weight reading for each perch, generated by a known small weight placed on the perch. This provides a base line for subsequent readings. The electronic system then runs unattended for the rest of the day, accumulating bird weights from any of the perches. It is not possible to ensure that only one bird is responsible for the reading at any weighing. When the birds are small two may be able to mount the perch at the same time, and at any stage of their development one of the birds surrounding the perch may interfere with it. This unreliability must be reduced as far as possible, so the computer is programmed to accept only those readings which lie within ± 10 g of the running average. At the end of the day the meter can be switched to show the tare weight, the running average, the number of birds counted and the sub-total in each of a set of weight steps above and below the average. If the system includes a VDU this can display a histogram, together with the mean and standard deviation of the measurements (plate 17).

Plate 17 Computer displaying histogram of bird weights, derived from strain-gauged perches (NIAE).

In broiler houses it has been found that cock and hen birds (which differ in weight) tend to have different perching habits. In principle, the computer can make a correction for this, given the cock/hen ratio of the crop, but even without this adjustment the system provides a close indication of bird weight. Calibrations against manual weighings have rarely shown differences between the averages (manual v. automatic) of more than 50 g.

6.5 Ultrasonic testing

Ultrasonic pulse-echo methods have been used for years for in vivo estimation of the amount of back fat on pigs, with an accuracy of about ± 10 mm. Increasing consumer preference for lean meat brings consequent economic advantage from production of leaner carcases, rather than those with waste fat on them. Back fat measurement is therefore particularly valuable to pig breeders.

The method and the equipment were originally devised for detection of flaws in metal structures and components. The sensor is a ceramic piezoelectric transducer which both generates and detects acoustic vibrations in the MHz range. Short bursts of these vibrations are transmitted by the transducer into the medium under test, by energising it at the appropriate frequency. It is then switched electronically to the 'receive' mode and subsequently detects the vibrational energy which returns by reflection at boundaries between zones in the material which have different acoustic properties. A series of such boundaries produces a succession of 'echoes', each delayed relative to the initial burst of ultrasonic energy, by a time dependent on the distance from the surface position of the transducer to the boundary concerned and by the velocity of the acoustic waves in the material.

In meat there is a sufficient distinction between fat and lean acoustically to create echoes. At 24°C the velocity of the acoustic waves in fat and lean (muscle) meat is about 1.45 km/s and 1.6 km/s, respectively. The electronics measurement system normally produces a picture on the screen of a cathode ray tube, depicting the transmitted pulse and the echoes, distributed along a horizontal timebase. Timing markers are superimposed on the display, to facilitate calculations of delay time and hence the depth of the fat/lean or lean/fat boundaries.

Normally, a 25 mm transducer probe is used and the pig's skin is shaved at the test point. A bland grease or oil is also applied to the skin, to provide the acoustic coupling needed for efficient transmission of the vibrations into the animal's tissue. Without this coupling much of the ultrasonic energy would be reflected at the surface of the skin. Application of the probe to the skin, and interpretation of the echo pattern on the screen, call for skilled operators. However, the function of interpretation can be taken over by a suitably programmed computer. This makes it possible to produce a two-dimensional picture of the lateral fat/lean distribution at a section of the pig's back, from a lateral scan by the transducer of that section. To achieve this the successive test positions of the transducer must be known, so the probe must be used with an external frame, which automatically provides the co-ordinates of each test position.

It has been found possible to perform similar measurements on lambs by employing a probe of 5 mm diameter and without shaving the animal at the test point. It is sufficient to part the fleece there and to use an oil coupling medium. Correlations between the estimates of fat and the measurements made on lamb carcases have been good.

The technique has also been employed to detect pregnancy in ewes under farm conditions. Non-pregnant ewes have been found to produce echoes at 60 mm to 80 mm from their skin surface, whereas pregnant animals show little or no response at this range, but produce echoes in the range 120 mm to 200 mm. Pregnancies can be detected with a high degree of certainty from about ninety days after service, thereby giving the farmer information on his ewes' fertility and enabling him to adjust their feed rations accordingly.

6.6 Egg grading

This operation is still done manually, in general, although interior defects of eggs, such as blood spots, have been detected automatically by sealing them in apertures in a moving belt which is strongly lit from below, and measuring the light transmitted through them with a sensitive photodetector system. By optically filtering out the amount of transmitted light in two adjacent, narrow wavebands (near 577 nm and 597 nm, respectively) and comparing the difference in the intensities with the aid of a differential amplifier circuit, the effect of differences in shell colour is reduced. Blood diffused in the egg's albumen can be detected at the 5 p.p.m. level and blood spots between 1.5 mm and 6 mm diameter in it can be detected with a reliability that increases to certainty at 6 mm diameter. However, the effectiveness of the method depends on exclusion of all light from the photodetectors except that transmitted through the eggs and this requires near perfect sealing of the aperture in the belt by each egg. Also, the measurement is slow — only about three eggs per second per channel.

Human graders, working with back-lit batches of eggs, can work much faster and are not so troubled by light leaks. Nevertheless, on some egg-grading lines manual removal of defective eggs from a travelling belt is being replaced by automatic means, employing a system originally developed as an aid to manual grading of potatoes. This semi-automatic system employs a matrix of sensing coils (fig. 6.4), fixed below the conveyor at the grading point, which match the row and lane spacings of the eggs on the conveyor. The graders wield a small actuator (dubbed 'magic wand', for obvious reasons) with which they lightly tap any defective egg, using the tip of the 'wand' for this purpose. The tip contains a piezoelectric crystal which generates a voltage when the egg is tapped. Thus triggered, the actuator radiates a signal which is picked up by the coils under the conveyor. The strongest signals are picked up by the nearest coils, which are identified by a comparator/decoding circuit, to provide the lateral and longitudinal co-ordinates of the defective egg, in relation to its position on the belt. As in vegetable and fruit grading, the circuit stores this information in memory and automatically relates it to the movement of the conveyor. The egg is subsequently discarded by a rejection mechanism when it reaches the discharge end of the line. The use of the system is not limited to grading on the basis of internal defects, of course.

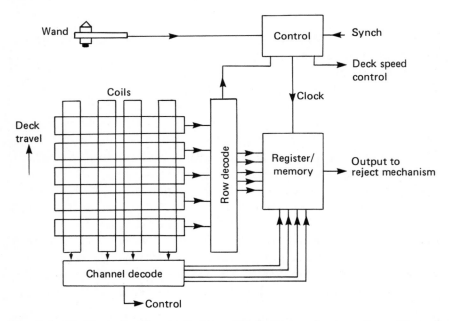

Fig. 6.4 'Wand' and coil matrix (simplified) for semi-automatic grading of produce on conveyor lines (SIAE).

6.7 Pig and poultry wastes

The introductory section to this chapter identified biogas production as an active area of development for microelectronic monitoring and control systems. In fact, production of this fuel is only one element in the more efficient utilisation of animal wastes and it brings problems of contaminants in the fuel, of matching production to requirements and costs of storage, if supply and demand are not equally matched. Nevertheless, on-farm digesters have been set up in larger pig and cattle enterprises and rising energy costs are likely to bring about an increase in their number.

The engineering requirements are to design a system which will run with the maximum reliability in farm operation, without the need for staff skilled in process control and with optimum efficiency, which includes minimising consumption of the fuel generated to maintain the temperature of digestion. In part, reliability is determined by the properties of the waste material put into the digester and this will be affected by the amount of preliminary filtering carried out, to remove indigestible material and farm debris of various kinds. For the rest, complete and automatic monitoring and control of the important process variables is essential and this makes the incorporation of a microprocessor-based system inevitable. In particular, it must monitor and control the temperature of the process, which is crucial, and the dwell-time of the material in the digester. There is scope for transducer development in this area, to improve the level of reliability and control (measurement of total and volatile solids, for example).

It is unlikely that similar applications of electronics will find a place in the poultry industry. Poultry manure is equally suitable for biogas production but the amount produced on most farms is small and, by virtue of its high nitrogen content, it is more valuable as a fertiliser or feed component.

6.8 Future developments

This sector of electronics applications has considerable potential for expansion within existing technology but the capabilities of electronic monitoring and control will be extended through instrumentation research, some of which will have an equal bearing on the cattle industry. Monitoring of the environment in pig and poultry houses will benefit from the introduction of a range of inexpensive and robust transducers which will monitor pollutants in the atmosphere and heat loss from the stock, directly or indirectly. There is room for improvement of gas concentration and egg weight monitoring in hatcheries, too, as noted in 6.2.2. Monitoring of eggs for the detection of internal and shell defects presents another challenge. Laboratory methods developed so far have required too much time per measurement to make them commercially viable.

Another possibility for the 1980s is the development of on-farm methods of feed analysis. These are discussed at a little more length in the next chapter, since they relate to on-farm production of cattle rations, but insofar as the poultry or pig man mixes his own rations the measuring equipment will be applicable to his sphere. Individual feeding of sows, on the lines of individual cow feeding, will provide a further application of miniature, implantable automatic coding devices, which will also facilitate other forms of performance monitoring and recording. Chapter 7 again provides more details.

Attempts have been made to develop fully automatic weighing systems for pigs, but these have not proved either safe or satisfactory. Nevertheless, some form of pen weighing may be evolved, since farmers often show an interest in this concept. If this is to be achieved, the location of the weigh platform will have to be based on the behavioural patterns of pigs and its design will have to be both simple and reliable in a difficult environment.

The need for new sensors in waste digestion plant was mentioned in 6.7. There seems a reasonable expectation that chemical and chemical engineering research centres will produce the equipment required.

Most of the new transducers mentioned above will take their place in integrated monitoring and control systems. For example, environment, feed and water intake of pigs and poultry should be monitored and controlled centrally, since they all contribute to the production process and all interact. Also, the central computer can merge the measured data with other data, to provide comprehensive management information, as in other sectors of agriculture.

6.9 Further reading

General

Sainsbury, D. (1980). *Poultry Health and Management*. St Albans, Herts: Granada Publishing Ltd.

Whittemore, C.T. (1980). *Pig Production. The Scientific and Practical Principles*. New York and Harlow, Essex: Longman.

Section 6.2

Bruce, J.M. and Clark, J.J. (1979). 'Models of heat production and critical temperature for growing pigs'. *Animal Production* **28**, 353-369.

Ministry of Agriculture, Fisheries and Food (1977). *Bulletin 148. Incubation and Hatchery Practice* 6th edn. London: Her Majesty's Stationery Office.

Randall, J.M. (1977). *Report No. 28. A Handbook on the Design of a Ventilation System for Livestock Buildings Using Step Control and Automatic Vents*. Silsoe, Beds: National Institute of Agricultural Engineering.

Walker, J.N. ed. (1974). *ASAE Special Publication SP-0174 Proceedings of Livestock Environment Symposium*. St Joseph, Michigan: American Society of Agricultural Engineers.

Section 6.3

Randall, M.J., Herbert, M.J. and Hepherd, R.Q. (1976). 'An automatic pneumatic conveying and weight-dispensing system for floor-feeding pellets to pigs'. *Journal of Agricultural Engineering Research* **21**, 233-245.

Section 6.4

Smith, R.A. and Turner, M.J.B. (1974). 'Electronic aids for use in fatstock weighing'. *Journal of Agricultural Engineering Research* **19**, 299-311.

Turner, M.J.B., Gurney, P., Benson, J.A. and Crowther, J.S.W. (1981). *DN 1031 Automatic Weighing of Broilers — Investigation into Perch Design*. Silsoe, Beds: National Institute of Agricultural Engineering.

Section 6.5

Allison, A.J. (1971). 'The use of an ultrasonic device for detection of pregnancy in the ewe'. *Proceedings of the New Zealand Society of Animal Production* **31**, 180-185.

Gooden, J.M., Beach, A.D. and Purchas, R.W. (1980). 'Measurement of subcutaneous backfat depth in live lambs with an ultrasonic probe'. *New Zealand Journal of Agricultural Engineering Research* **23**, 161-165.

Johnson, E.K., Hiner, R.L., Alsmeyer, R.H., Campbell, L.E., Platt, W.T. and Webb, J.C. (1964). 'Ultrasonic pulse-echo measurement of livestock physical composition'. *Transactions of the American Society of Agricultural Engineers* 7, 246-249.
Lindahl, I.L. (1972). 'Early pregnancy detection in ewes by intra-rectal reflection echo ultrasound'. *Journal of Animal Science* 34, 772-775.

Section 6.6
Norris, K.H. and Rowan, J.D. (1962). 'Automatic detection of blood in eggs'. *Agricultural Engineering* 43, 154-159.

Section 6.7
Hobson, P.N. and Robertson, A.M. (1977). *Waste Treatment in Agriculture*. London: Applied Science Publishers Ltd.
Stafford, D.A., Hawkes, D.L. and Horton, R. (1980). *Methane Production from Waste Organic Matter*. Boca Raton, Florida: CRC Press Inc.

7 Cattle

7.1 Introduction

It was noted in the introduction to Chapter 6 that the beef cattle industry makes relatively little use of electronics, whereas the dairy industry is a major user. These two sectors are different in several ways, although they have in common low-cost, naturally-ventilated housing and the production of large quantities of farmyard manure or slurry, which may have to be treated before disposal. The beef animal normally lives less than two years before it is slaughtered: the dairy cow may live over three times as long, producing milk and calves for most of the time, before being culled from the herd. Unless they are out for summer grazing, both receive bulk rations which are fed to them, or to which they can gain access, while they are grouped in yards or housing. However, the beef animal may not enjoy the same standard of living as the calf-bearing, lactating cow, except when it is being 'finished' for market. Neither, as a rule, does it receive the 'personal' concentrate ration that is tuned to the needs of the modern high-yielding cow. A herd of the latter valuable creatures must have a modern dairy parlour, too, fulfilling all the requirements for hygiene that any milk producer must meet. The dairyman — usually among the best paid of farm workers, because so much depends upon his skill and stockmanship — also needs a reasonable standard of environment in the dairy. In fact, apart from some fan-ventilated calf houses in older buildings, environmental control otherwise plays little part in the cattle sector.

Overall, the dairy industry is exceptional in the livestock sector, because of its high capital investment and the precise and demanding management skills required of the farmer and stockman to achieve high outputs at minimum cost. Most of this chapter is about dairy herds, therefore, and the ways in which electronics can support people in this sector. References are made to beef cattle where the electronic systems described have applications to them, too.

The next section (7.2) deals with a topic which has a bearing on every aspect of electronics applied to larger dairy herds, defined as seventy cows or more. Herds of this size are becoming more common, and, in the U.K. at least, produce the majority of the nation's milk. There is little doubt that when the herd reaches this size it becomes difficult for the dairyman to remember and record all the important facts about each cow at each milking and at other times when she needs treatment of one kind or another. The problem becomes more acute if the regular dairyman is absent for any reason and a relief worker is in charge of milking operations. One solution to this problem is provided by the automatic identification systems which have become widely established on dairy farms in several countries. These systems are developing rapidly but a description of some of the pioneering designs should serve to illustrate their potential.

Measurement and control of the feed input to dairy cattle is important, although feed does not account for as high a proportion of the production costs as it does in the pig and poultry industries. Electronics plays little or no part in the feeding out of bulk rations of hay, although it does provide an input at harvesting, as described in 3.2.4. On the other hand, it has established a significant role in the feeding of silage and mixed bulk rations of other kinds, as described in the third section. This section also deals with the delivery of individual concentrate rations to the dairy cow.

Electronic weighing of beef cattle is as important as it is for pigs. The need for regular cow weighing is often disputed but many dairy farmers rate a knowledge of the weight trend of each cow as an important index of her well-being. The section on weighing therefore deals with both beef and dairy cattle, including a system of fully automatic weighing as well as the more usual semi-automatic method.

This chapter also looks at several aspects of milk harvesting and milk analysis. As elsewhere in this book topics outside the farm gate (in this instance, beyond the bulk tank) are only introduced if there appears to be a real prospect that the equipment concerned may have a parallel on the farm within the next decade or so. First, however, a section deals with the regular on-farm operation of milk yield recording, which concerns both the farmer and milk marketing organisation. Milk sampling and milk quality require some discussion before space is devoted to two topics which are more contentious (from the scientific standpoint) in the present state of knowledge. Certainly, many farmers are concerned about missed 'heats' and failed inseminations in cows, with their consequent financial loss, therefore they are interested in early detection of oestrus and pregnancy. Most farmers are even more concerned at the incidence of that expensive scourge, mastitis. There are prospects that reliable detection of oestrus and mastitis (in its clinical form, at least) will prove possible through suitable forms of milk monitoring.

Something is needed to bring together all the monitoring and control functions which centre on the dairy parlour but extend into nearby cattle houses and yards, with their feeding equipment. The 'something' is almost inevitably a micro- or mini-computer, based on microprocessors. A section is devoted to this topic.

The subject of livestock waste has to be raised in the present chapter, too, if only briefly, since it has been mentioned in chapter 6.

7.2 Automatic identification

7.2.1 Tuned coils

Equipment for automatic identification of cattle was first developed for nutritional scientists, who wished to feed individual rations to experimental animals in small groups of about twenty, and without a great dependence on labour. A successful method of doing this was evolved in the late 1960s and has been widely used since then. In essence, the system works in the following way. Each animal has a transponder hung round its neck by a plastic chain or neckband. The transponder has no built-in power supply, i.e. it is a passive device. In fact, the sealed, plastic transponder case contains an electrical coil with a tuning capacitor in parallel with it, and nothing more except waterproofing material. Each transponder can be tuned to a different frequency in the upper audio frequency — low radio frequency band (up to 100 kHz) by suitable choice of the capacitors. Given good design, which results in sharp tuning, it is possible to distinguish reliably between coils tuned to frequencies 2 kHz to 3 kHz apart in this waveband, even allowing for temperature drifts and ageing effects. Therefore, a small herd can easily be equipped with transponders, all of different frequencies but contained within a frequency range below 100 kHz. Every coil acts as a 'key' to one of a set of gates or other barriers, each of which leads to a feeding point. A cow is able to reach her individual ration only if she presents herself at the correct barrier. This barrier 'recognises' her because it contains an oscillator tuned to the same frequency as her transponder. As she brings the transponder towards the barrier it absorbs an increasing amount of power from the oscillator circuit, until the effect on the circuit is sufficient to trigger a recognition response. This actuates the electrical solenoid which unlocks the barrier and allows the cow to push through to her ration. The equipment can be used with beef animals too, of course.

The limitation on the number of animals that could be identified in this way led to a double tuned coil system intended for large dairy herds. The cow's passive transponder carried two coils, each of which was tuned to one of twenty frequencies. The recognition circuit cyclically radiated power at each of the twenty frequencies in turn. At close range, enough of this power was absorbed by the two coils at their tuned frequencies to operate a small radio transmitter in the transponder. The recognition circuit received the two transmissions and in this way determined which two out of the twenty frequencies were being picked up by the transponder. A combination of two out of twenty provided one hundred and ninety identities. The power absorption and retransmission system also enabled recognition to be carried out at a greater range. Large transmission coils, mounted on a walk-through archway, made it possible to identify a cow as she entered or left a dairy parlour or any other location, as well as functioning as a feeder access control. In the 1970s equipment of this type met the stringent requirements of the U.K. Home Office Radio Regulatory Department in respect of transmitted frequency spectrum and stability and low radiated power. Unfortunately, production of tuned coils to the frequency tolerances required proved to be difficult but the successful commercial designs which emerged in the mid-1970s adopted the power absorption and retransmission method.

7.2.2 *Pulse coding*

The system now widely used in the U.K. employs PCM (pulse code modulation) to indicate a cow's identity. The essential features of the system are shown in fig. 7.1. The energising coil on the left of the diagram is part of a power oscillator circuit, operating at a fixed frequency in this case. The oscillator is crystal-controlled, for frequency stability, and generates power at about 58 kHz. To meet Home Office standards its output is limited to 1W.

The animal transponder contains a ferrite rod aerial which picks up the power (about 4 mW) needed to operate the PCM circuit. The circuit employs mainly CMOS technology (2.2.1) with low power requirements and radiates serially a built-in digital code at a precise radio frequency (26.995 MHz) and about 0.5 mW power to the recognition circuit on the right. In effect, the '0's and '1's of the code word are produced by switching the r.f. on and off in the correct sequence. An 8-bit word will provide 256 identities but in view of the growing number of herds over 300 animals a double (16-bit) code word is employed. This provides 65 536 identities, which is more than enough for one farm but allows leading or trailing digits to be used for other information.

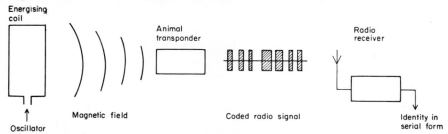

Fig. 7.1 Animal identification system using a transponder (NIAE).

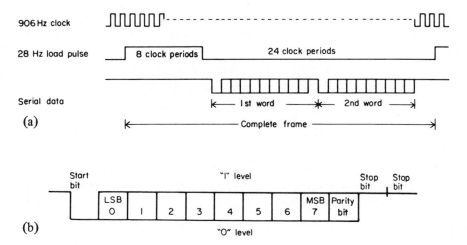

Fig. 7.2 Digital transmission of a 16-bit animal code number. (a) Transmission format and timing (b) Word format (NIAE).

This equipment provides a practical example of digital data transmission in the agricultural context, therefore its operation is described here in a little more detail. Within the transponder the 58 kHz received power is in part converted to d.c., to provide circuit power, and in part used to provide 906 Hz clock pulses, generated by binary division circuits. Further binary division to 28.3 Hz provides transmission control pulses. The data transmission sequence is given in fig. 7.2(a). The 28.3 Hz pulses divide the 906 Hz clock pulses into groups (frames) of 32; of which 24 provide the cow's 16-bit identity number, with a '0' gap between the first and second 8-bit words. Fig. 7.2(b) gives more detail of the bit-pattern of each word. Each starts with a '0', then follows with the 8-bit word which is programmed into the circuit, starting with the least significant bit (LSB) and ending with the most significant bit (MSB). The last bit of the word — the parity bit — is made a '0' or a '1' depending on the code sequence which precedes it, to produce an even or odd number of '1's, as required. In this case the two words which make up the 16-bit code have different parities — one even and one odd — to distinguish them. This is one of the safety measures employed in digital data transmission (appendix 2). If the receiving circuit does not register an even number of '1's in each even-parity group or an odd number in an odd-parity group it rejects the received data as corrupted. At the receiving end, a serial communications decoder chip collects the serial PCM data from the radio receiver, together with an input from the 58 kHz transmitter, which provides a reference signal. The decoder has 16 digital outputs which indicate the animal's code in parallel form.

In the U.S.A. a higher power, microwave frequency system (915 MHz) was developed in the 1970s for longer-range detection of cattle outdoors as well as for dairy units. This employs a small, passive transponder which can be implanted. It has been placed under the animal's skin near the backbone. The small transmitter receiver, with its equally small microwave aerial, can interrogate the transponder at up to 1 m if it is implanted and up to 15 m if it is external to the animal. The transponder replies with a 16-bit identity number and another 16-bit word which reports the animal's sub-dermal temperature, measured by a sensor on the transponder, if the latter is implanted.

Details of the applications of these systems are given in following sections.

7.3 Feeding

Feeding of cattle can be considered under two broad headings, namely bulk rations and concentrates. The former topic can be further divided, according to the method of distribution. Conveyor

systems and feeder wagons provide two distinct methods of bulk
feed delivery which are treated separately here.

7.3.1 *Conveyor systems*

Belt or chain-and-flight conveyors are in use for distribution of bulk
rations of silage, alone or with additives, to groups of cattle. Normally
− although not necessarily − they are associated with tower silos.
The uniformity with which rations can be dispensed by these means
is generally within acceptable limits. Electronic weighing has been
introduced to control the total amount of silage dispensed to groups
of cattle at each feeding. The weigher − usually of the continuous
type − is placed between the silo unloader and the conveyor lines.

Several types of continuous weigher have been marketed for use
with silage. This is a difficult material to handle, capable of corroding
unprotected metal parts very rapidly, and the design of the weigher
must be such that silage cannot cling to and build up on moving
parts. Apart from this mechanical aspect of the design there is an
electronics problem, too. Because of the compressed state of the
material in silos the cutting and unloading mechanisms tend to
produce an irregular output flow, which can be reduced to almost
nothing from time to time. In consequence, the weigher must be
sensitive to quite small amounts of material and at the same time the
drift in its output with time must be low. Its ability to meet these
requirements has a considerable influence on the accuracy with
which the total mass flow over it can be measured.

The weigher to be described here (plate 18) possesses the required
characteristics and has the additional feature that it will control the
proportion of additives in a silage-based mixed ration.

Mechanically, the machine contains an endless, reinforced plastic
belt, about 0.6 m wide, which runs over two rollers with centres
2.4 m apart, supported in a rigid frame. The top run of the belt
slides over three continuous platforms between the rollers. One platform at each end supports the belt and the material on it as it enters
and leaves the weigher (i.e. the lead-on and lead-off platforms,
respectively). The centre section is the weighing platform. Side
guards retain any silage which strays to the sides of the belt and
rubber strips attached to each guard rest on the edges of the belt,
to protect them. Any silage which does find its way under the strips
falls through side slots onto the ground. A transverse nylon brush
below the front roller prevents silage build-up on the belt itself. The

frame assembly is connected to a base frame at three points — a design which avoids distortions of the upper assembly if the machine is used on uneven ground. The belt rollers are tapered slightly at each end, to assist correct tracking of the belt. Belt tension is controlled by springs on the rear idling roller.

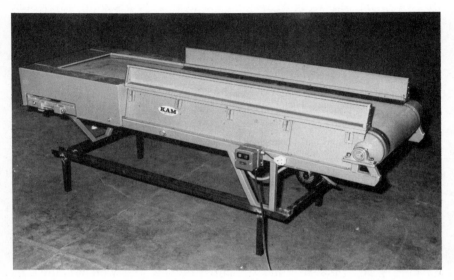

Plate 18 Continuous belt weigher for silage (Knee Agricultural Machinery Ltd.).

Electrically, the belt is driven through the front roller by a 745 W electric motor, via a reduction gear drive, at a speed of about 0.75 m/s. The weighing platform is connected to the lead-on platform by hinges and is supported at the lead-off end by a load cell. This transducer has a load limit of about 11 kg, with a corresponding displacement of 0.13 mm, and a temperature coefficient of 0.005% (of its F.R.O.) per °C. The three platforms are carefully aligned to be co-planar at no load and when this is done the 0.13 mm deflection of the load cell under maximum load has little effect on the accuracy of weighing.

A block diagram of the electronics circuit is shown in fig. 7.3(a). The amplified signal from the load cell passes through an active filter circuit which has a low frequency pass-band (up to 1 Hz) and 12 dB/octave cut-off beyond that frequency. This removes signals due to vibration. The auto-zero circuit which is interposed between the filter and one input to a differential amplifier helps to reduce the effect of drift in the unladen output from the weigher. This circuit samples the output from the filter when the ZERO push-button is pressed momentarily, the duration of the sample covering one complete revolution of the belt. The zeroing operation is performed just before weighing begins and the sampled filter output provides an average of the zero weight signal, which includes the effect of the join in the belt. The average signal level is retained throughout the weighing by a 'hold' circuit and supplies the differential amplifier with a zero weight reference. The conditioned output from the load cell provides the second input to the amplifier, which therefore produces a signal proportional to the mass of silage on the weighing platform at any moment.

The remainder of the circuit processes the instantaneous weight signal. A voltage/frequency converter changes it to a pulse train which is accumulated in a digital counter. The count is continuously compared with the manual setting of the amount of silage required by the animals and when the two numbers are the same an output from the comparator circuit terminates the feed cycle. The output from the differential amplifier is also taken to the proportioning circuits which control the addition of minerals, barley, etc., to the silage in proportion to its instantaneous mass flow. These circuits are shown schematically in fig. 7.3(b). The thumbwheel switch for manual setting of the silage ration to be dispensed also adjusts the voltage gain of amplifier A1 inversely (i.e. the larger the amount of silage the smaller the gain). A second switch, for setting in the amount of an additive required, determines the gain of amplifier A2 proportionately. Therefore, when the silage flow signal is applied to the input of A1, the output from A2 depends on both the flow and the preset ratio of additives to silage. As shown in fig. 7.3(b) this output signal is employed to control an auger, which spreads the additive on the silage at the required rate. The output from A1 can be coupled to similar circuits for proportional addition of other components of the ration.

A weigher of this type has been used to measure flow rates of forage up to 25 t/h but in more normal usage it has operated long-term at rates between 1 t/h and 6 t/h, with a coefficient of variation (1.2.2) of about 2%. Farm trials of the proportioning system at different silage flow rates, using rolled barley as the additive, showed that the required proportions can be met with a coefficient of variation less than 2%.

Calibration of the weigher on the farm can be done by passing a known weight of material (say, grain) over it at a steady rate but it is equally effective to pass a 2 kg weight over the weighing platform repeatedly, as quickly as possible. This can simulate flows up to about 10 t/h and a reliable calibration takes only a few minutes.

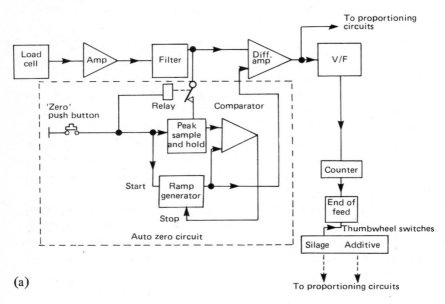

(a)

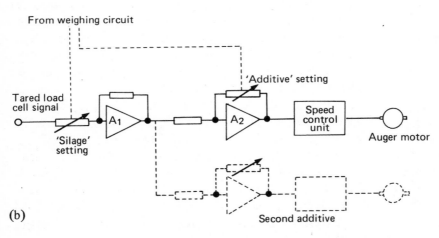

(b)

Fig. 7.3 Continuous belt weigher for silage. (a) Weighing and recording circuit (b) Proportioning circuit for additives (NIAE).

The silage weigher does not work alone. It is linked with the silo unloader, elevators, conveyors and feed divertors of various kinds which must be synchronised and monitored if feed distribution is to be largely automatic — as it can be, given well designed and properly maintained mechanical equipment. The complete conveyor feeding system therefore requires a central controller, which starts and stops the sequential elements in the system at the proper times, or takes emergency action (alarm bells and shut down) if a monitoring device indicates a fault condition.

As an example of normal timing requirements, when the continuous weigher produces the 'end of feed' signal the silage still in the feed distribution line must be cleared before the whole system is shut down; otherwise deteriorating material will be left to cause problems at the next feed. Material already cut from the silo and upstream of the weigher continues to pass over it after the 'end of feed' signal for a time, and some allowance must be made for this if maximum precision of feeding is required.

Emergency conditions are normally signalled by flow monitors placed at strategic points in the system. These may be of the rotating paddle type or may be simple flap-operated switches. Their indications are ignored by the controller for a short while after start-up of feeding but alarm action is taken if a no-flow state continues, or develops during the course of the feeding programme.

In larger conveyor systems, the controller may feed several groups of cows in sequence, each group having a different mixed ration. In this application it will follow a preset programme, automatically selecting the proportions of the components in each feed and routing it through the conveyor lines to the required destination. If soundly designed, these installations are very reliable.

7.3.2 Feeder wagons

Many farmers prefer a mobile feeder wagon or forage box to the fixed conveyor system, particularly because their silage may be stored in a bunker (horizontal) silo at a considerable remove from the cattle feeding houses or yards. These mobile dispensers dispense forage- or treated straw-based rations from the side or rear as they are driven or hauled past the cattle feeding troughs.

The mixer/feeder wagon fitted with an electronic weighing system was pioneered in the U.S.A. early in the 1970s and has found considerable favour elsewhere since then. The types and positioning of the load cells vary from model to model but basically the body of the wagon is supported on three or four of these cells, which are in turn supported on the wagon's chassis. The output from the cells is summated and the total load is displayed digitally by the associated control box. The box contains sets of digital switches which are used to set up the required amounts (in kg) of each ingredient in a mixed feed. When this has been done filling of the wagon can start and the operator continues loading the first ingredient (using a fork loader, auger, etc., as appropriate) until the control box produces a warning light to indicate that the required amount of this material is in the wagon. The operator then loads the second ingredient until the required additional weight is reached, at which point the warning light (which switches off during the second loading) reappears. The sequence continues until all the ingredients have been added. After the contents have been mixed the wagon is taken to the discharge point and the rations are dispensed as uniformly as the unloading mechanism allows, using the warning light to indicate when a preset amount has been discharged.

The cumulative method of filling the wagon, just described, is capable of an accuracy of ± 5% or better but it is liable to greater error if the operator is unable to load the wagon with the required degree of precision. The man filling the wagon with a front loader is usually not in a position to observe the meter's display very clearly and must await the warning signal from the light before ceasing to load a feed component. It is very easy for him to dump a large forkful of material into the wagon and in so doing to exceed the preset weight by a considerable margin. A study of the operation of one fork loader showed that its load could be anything from 80 kg to 150 kg at a time. When this happens the next component will be underweight by about the same amount unless the operator reschedules the control programme.

One way to control this is to set the programme so that another warning light appears, say, 50 kg to 100 kg before the required amount of an ingredient has been loaded. However, an advance on this is the auto-taring system which takes the actual weight of the first ingredient loaded as the baseline weight for the second component, and so on. In use, the control box allows the operator to dial in the required weight of each ingredient (up to four ingredients, say) at the start. He then loads in the normal way, being warned by an amber light when he is at 90% of the target weight for the ingredient concerned, and being given the red (stop) light at the preset weight. At this point the circuit waits five seconds, to allow any vibrations to cease and to give the operator time to check that the

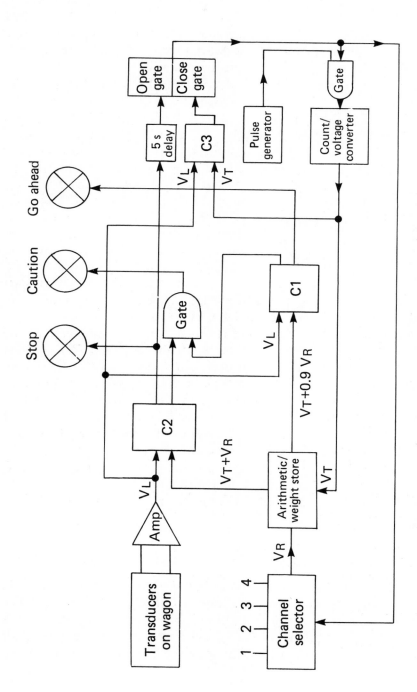

Fig. 7.4 An auto-taring system for feeder wagons, with 'traffic lights' indicators (NIAE).

loading equipment is not interfering with the weighing. Then the controller auto-tares the load and steps on to the next channel. Unloading follows a similar pattern. Apart from the greater control over the amount of individual ingredients that auto-taring brings, it is also better for the driver, who is not required to leave the tractor cab to change the settings of the controller for any reason.

Figure 7.4 is a simplified block diagram of the auto-taring circuit, showing the logical sequence of operations. At the start, with the wagon empty, the amplified signal from the load cells, V_L, is entered as the first tare voltage, V_T, in the reference weight store. The target weights of the ingredients (up to four in this case) are entered manually and a voltage, V_R, representing the first of these, is transferred to the reference store. Voltages $V_T + 0.9 \, V_R$ and $V_T + V_R$ appear at the inputs to two voltage comparators C1 and C2. Because at the start V_L is less than $V_T + 0.9 \, V_R$, C1 operates the green 'go ahead' lamp. As filling of the first ingredient proceeds V_L increases and as soon as V_L is greater than $V_T + 0.9 \, V_R$ C1 switches over, the green lamp is extinguished and the amber 'caution' lamp is lit. At V_L equal to or greater than $V_T + V_R$ the amber lamp is also extinguished and the operator sees the red 'stop' sign. After the five seconds' delay already referred to the actual value of V_L is stored as the new V_T, together with a new V_R, representing the required amount of the second ingredient. The programme continues as before. This outline of the operation of the circuit has been given in terms of analogue voltages and, in fact, the prototype auto-taring system employed a mixture of analogue components and digital logic. Nevertheless, apart from the transducers and their associated amplifier, the required circuit functions – storage, comparisons and switching of lights – are clearly ideal for a microprocessor and memory system.

Inherently, the weighing system possesses ± 1% accuracy or better; the actual accuracy obtained under farm conditions still depends on the care with which the operator loads the last 10% of each ingredient.

7.3.3 *Concentrate dispensing*

Some dairy farmers feed individual concentrate rations to their cows in the dairy parlour; others believe in out-of-parlour feeding. The electronics engineer has no need to take sides, fortunately: electronics can contribute to efficient use of concentrates in either case. Whether she is in or out of the parlour, the presence and identity of a cow can be automatically detected at a feeding point and her ration delivered to her via an electrically-operated concentrate dispenser. Apart from the consequent labour-saving, data on her actual feed consumption at any time can be gathered, as an aid to management.

One pioneering out-of-parlour system employs a cow transponder which collects radiated power from an external high-frequency source to energise a retransmitting circuit, like the transponders described in 7.2. It differs in principle from the latter, however, since it does not carry a code number. All transponders of this type contain a retransmitting oscillator with the same frequency. When a cow wearing the transponder round her neck approaches a feeding point the power source energises the oscillator, which retransmits to a nearby radio receiver. The output from the receiver starts a feed auger which begins to dispense her ration at a fixed rate. At the same time the transponder employs some of the energy that it receives from the external source to charge a small, internal, rechargeable battery at a constant rate. The cow can eat until she is satisfied or until the battery is fully charged, at which point the retransmitting oscillator is switched off and feed dispensing stops. The transponder battery also discharges at a linear rate through another part of the circuit and is effectively completely discharged after twelve hours, unless the cow returns for more feed. By this means the cow can eat little and often or more substantially at longer intervals. The rate of charge can be adjusted to control the amount of feed that she is given in a twelve-hour period and the twelve-hour discharge period also limits the amount that can be dispensed to her at one feed. Tests of this system showed that the cow's concentrate ration can be controlled within ± 10%, with a 99% level of confidence.

The automatic identification transponders of 7.2 are employed in other concentrate dispensing systems for in- or out-of-parlour feeding of 250 or more cows. When the recognition circuit has picked up the cow's identity this is referred to the corresponding address in a semi-conductor memory, in which is stored the amount of her daily ration, preset via a keyboard. This information is passed on automatically to the controlled dispenser, which feeds her a predetermined part of the ration. Out of parlour, this operation is subject to overriding control by another preset programme, which determines the permitted time between meals. These programmable feeders are often used with electrically-operated auger dispensers, and sometimes with vibratory feeders. However, for greatest accuracy a gravimetric dispenser is to be preferred because it is less affected by physical differences (particle size, moisture content, etc.) between feeds. Coefficients of variation of less than ± 2% are to be expected from the gravimetric dispenser, compared with at least twice that figure for other types.

Most manufacturers of programmable out-of-parlour feeders provide one central controller which can operate multiple feeders 100 m or more away if required (but note, long cables are costly). All but the simplest contain microprocessors, which enable the system to record the actual daily feed consumption of each cow and to increase the next day's ration of any animal that does not take up all of its ration on a given day. The system will also draw attention to any individual failing to take up a significant percentage of her ration for any reason. Output is by alphanumeric display and/or printer. Nearly all have stand-by battery supplies which will sustain the system for at least one day if the mains power fails.

Successful unattended operation of these feeders depends on the design of the feeding unit: dominant or voracious cows can prevent others from eating unless some form of stall protection is afforded to the animal at the feed trough. Figure 7.5 shows the feeder and stall in a commercial unit. Limitation of the amount of concentrate dispensed at one time makes it unlikely that any cow will stay at the feeder for more than a few minutes and the record of feeding patterns provided by the electronic equipment makes it possible to identify individual cow problems.

7.3.4 *Feed analysis*

In forage analysis the proportion of a constituent is normally reported in terms of the total dry matter (DM) of the forage, because this gives the required indication of the material's feed value. Hence protein content is quoted in g/kg DM and metabolisable energy (ME) is MJ/kg DM. Since most forage is grown on the farm where it is consumed, on-farm analysis by electronic means — akin to grain protein determination (5.3.6) — is one goal of agricultural engineering research. Measurement of dry matter is clearly fundamental to progress in this direction. Fortunately, two methods of moisture determination described in 3.4.1 (infra-red reflectance and microwave absorption) provide ways of making this measurement. Rapid sample measurements will enable the farmer to regulate each feed allocated to his cattle according to its dry weight. On-line moisture measurement in conveyor systems is a possibility, too, for automatic feeding to dry weight rather than total mass.

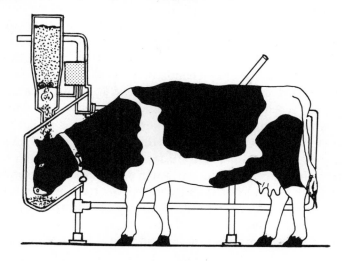

Fig. 7.5 Automatically-controlled out-of-parlour feeder. Cow with identification collar, in stall (**Alfa-Laval Ltd.**).

7.4 Weighing

7.4.1 *Semi-automatic weighing*

The semi-automatic pig weighing system described in 6.4.1 led to the development of similar equipment for weighing beef and dairy cattle. When cattle weighing is done manually one of the two or more operators opens and closes the front and rear gates of the weight crate, not without personal risk at times, and reads the weight, leaving someone else to line up the animals for passage through the weigher. Semi-automatic weighing is not reduced to a one-man operation as it is with pigs: it still requires a man bringing up the animals and another operating the pneumatic gate controls, as well as reading the weights. The benefits of the method are a smoother, quicker flow of animals through the weighing system and the avoidance of risk to the man operating the weigh crate controls.

The electronic measurement system developed for pig weighing requires little adaptation for cattle. Apart from the obvious need for it to cover a very different weight range, a load cell transducer is more suitable than the linear displacement transducer employed with pigs, since it operates with little movement of the weigh platform and many cattle are less reluctant to enter the crate in consequence. However, the subsequent signal processing, averaging and presentation of a steady, digital reading of the animal's weight is the same for pigs or cattle. The electronics design is therefore relatively straightforward but, as in so many other applications, the effectiveness of the system depends on the reliability of the equipment which interfaces with the outside world. In this case the world is represented by large and sometimes very lively beasts, which have to be contained without damage to themselves or the weigh crate for the few seconds required to take a weighing. To achieve this under pneumatic control the inlet and outlet gates of the weigher must be redesigned. The inlet gate, in particular, must be strong enough to close behind an animal but it must not subject the beast to pressure or impacts which endanger it. One satisfactory design employs two hock bars, one above the other, connected by vertical, pivoting straps. The two bars are also pivoted at one side of the gate, which opens vertically in the manner of a railway crossing barrier, under the control of a pneumatic ram. The whole assembly is partially counterbalanced. In the closed position the free ends of the two hock bars rest in catches, which oppose any backward thrust by the animal and the upper bar is held by a spring-loaded latch, which opposes vertical thrusts on the gate. The latch is opened pneumatically when it is required to open the gate. By this means the cattle in the 100 kg to 500 kg weight range have been weighed in a crate of average size, without difficulties arising from following beasts making attempts to climb over or squeeze under the closed gate. Another design that has worked well in practice employs a rising and falling gate with top and bottom crossbars and vertical (fixed) ties, which is mounted on two long arms, pivoted at the exit end of the crate. Counterbalance is provided by a pair of constant tension springs, one on each side of the crate. In the raised position these gates provide the clear view into the crate, which encourages the more timid beasts.

The exit gate should also provide a clear view ahead but must prevent an animal from attempting to jump through any gap in it. A close-spaced steel mesh gate, with an outward-facing mesh cage for the animal's head, has been employed successfully. Like the inlet gates above it is controlled by a pneumatic ram and fitted with a spring-loaded latch with its own small ram. Otherwise, a standard exit gate with head yoke has been employed but additional pneumatic operation of the yoke is required if the required smoothness and speed of operation is to be attained.

Adjustable internal side walls also assist the operation with smaller animals. These walls are arranged to taper inwards towards the exit end of the crate.

Given attention to this kind of detail the animals can be passed through the weigh crate at a rate of about 150/h, with minimum hazard to themselves and the operators. Overall, allowing for the time needed to collect and return them from their pens, throughputs of 80/h are possible.

Weighing accuracy depends on levelling the weigher, which should be done to within 1° from the horizontal for the best performance. The load cell should be protected from thermal gradients caused by wind gusts, too. A foam plastic wrapping, protected by a waterproof film, is effective in this respect. With these precautions a 400 kg animal can be weighed to better than ± 1% accuracy. Check weighings can be made with known weights of various kinds, including people.

It must not be forgotten, however, that the body weights of cattle vary more than 1% diurnally. In fact, a beast may lose several kilograms by defaecation or urination at any time. This variation can be minimised by weighing the animals at set times in the day but day-to-day the coefficient of variation can easily be 1%.

7.4.2 *Walk-through weighing*

The difficulties of unattended weighing have been noted earlier but in the case of the dairy cow it has proved possible to do this without risk to the animals, although at some cost in complexity and in the reliability of the readings. The ability to weigh the dairy cow in this way arises from its generally placid behaviour and the regular pattern of movement associated with milking. A walk-through weigher, placed in a passageway beyond the exit from the dairy parlour, has obvious attractions. By this means weighing is always done at about the same times each day, after all the cows have been milked out and without calling on busy dairy parlour operators. Furthermore, they can be weighed as they walk, without hindrance, along the familiar route to the dispersal yard.

The first requirement is clearly for an automatic means of identifying each animal, since they are not within range of the dairy workers. The radio frequency transponder systems described in 7.2.2 provide the equipment required for this purpose. An archway coil, mounted at the exit from the weigher, energises the transponder just before the cow leaves the weigher and an adjacent receiver detects her identity. A multi-core cable conveys power and the returning identification and weight signals to the central monitoring point in the dairy. The cow is identified at the exit rather than the entry because at this point all of her weight is on the platform and it is easier to relate her identity and weight in the subsequent data processing. Failure to register an identity results in the rejection of the weight data.

Plate 19 Walk-through weigher for dairy cows, with strain-gauge load cell (NIAE).

Plate 19 shows a walk-through weigher placed in the exit race from a parlour. A standard commercial weigher has been adapted for this purpose and fitted with a tension link load cell instead of a spring balance. This provides the firm platform which does not disturb the cows, as already noted. The length of the platform represents a compromise. If it were long enough for the animal to remain on it for two or three seconds it would provide the accuracy obtainable with the gated weighers described earlier. Unfortunately, since cows may bunch on leaving the parlour a long platform brings a high risk that more than one animal may be on it at one time. The weigher shown in plate 19 has the floor extended from 2.1 m to 2.5 m. This does not eliminate the possibility that a cow will step on the platform before the animal ahead has left the weigher, but it does reduce the frequency of this occurrence. Since it is not possible to rely on separation of the weight recordings of the cows the associated electronics circuits have to perform two functions, apart from identification, namely:

(i) Calculation of a weight from the dynamic load cell signal which it associates with each identification;
(ii) Rejection of any calculated weight which is invalid because the cow has been jostled, followed too closely by another, or has merely hurried over the platform in such a way that the true peak signal was not reached.

The first of these functions is performed by a filtering and peak-hold circuit. This operates on a dynamic weight signal which can vary in form, as shown in fig. 7.6. As would be expected, during the first half of the weighing the signal reaches a peak as the cow moves fully onto the platform and the second half brings a return to zero as soon as she leaves it. However, the dynamics of cow and weigher are such that the signal has two fairly well-defined signal levels, the higher one providing the required information on the animal's weight. The minor fluctuations of the signal are filtered out before the signal is passed to the peak-hold stage, which extracts this information for display and/or print-out of the weight in kg, if the measurement is passed as valid.

The second function is performed by a microcomputer which builds up a running average of each cow's weight from the value logged at each milking. The computer is programmed to reject any measurement which is more than ± 30 kg from the calculated average. This figure has been derived from analysis of selected records, which showed that although the running average of a cow's weight may change very little over several weeks individual weighings may differ from the average by ± 25 kg. The ± 30 kg limit is therefore as close as it can be in practice. The running average for any new cow in the milking herd can be started from an assured value by halting the animal in the weigher manually after its first milking while the computer registers the load cell's output. Subsequently, at the beginning of each week the value of the running average is stored and treated at the first weighing of the next week's sequence.

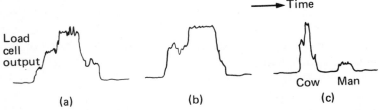

Fig. 7.6 Dynamic weight records of a cow moving through a walk-through weigher. (a) Cow rubbing its neck against the weigher, recorded weight 612 kg (b) Clear passage, with rapid exit (604 kg recorded) (c) Cow chased through the weigher (585 kg recorded) (NIAE).

Experience with walk-through weighers so far has shown that the computer may reject between 40% and 50% of the weighings at each milking, on the ± 30 kg test. This may seem high − and unusually lively or gregarious cows produce fewer valid readings than others − but in general a week's records can be expected to provide information on an individual cow's body weight trend similar to that obtainable from weekly manual weighings. Apart from this, the design and siting of the approach passage to the weigher have a substantial influence on the percentage of valid readings taken and there are expectations that through further study of these factors higher acceptance rates will be attained. Therefore the walk-through weighing system already provides useful indications of body weight trends, with prospects of improvement. Apart from their value as a health indicator, the weight trends can be used by the farmer to decide on individual rations and, in combination with measured milk yields, on the time for disposal of an animal.

7.5 Milk yield and quality

In the early 1980s both the milk recording jar and the pipeline flowmeter find a place for measurement of individual milk yields. Each has its particular merits and its adherents. As with concentrate dispensing, so with milk yield recording; the electronics engineer can remain impartial. Either can be fitted into any automatic monitoring and control system for the dairy parlour, in combination with automatic cow identification and the monitoring systems described in 7.7. This section deals with each in turn before moving on to metering of milk from the refrigerated bulk bank in the dairy to the milk tanker which transports it to the distribution centre. Although there is little on-farm involvement of electronics in milk sampling and analysis, this subject is reviewed briefly. Finally, reference is made to a topic related to parlour milk recording, namely automatic cluster removal.

7.5.1 Recording jars

Taking these first, there is one obvious way to measure a cow's yield automatically and another, less obvious one. The first is to mimic the operator, by measuring the level of the milk in the jar at the end of milking and the second is to weigh the jar. The latter may appear at first sight to be less obvious because of the attached flexible tubes which lead in from the milking-cluster on the cow and lead out to the vacuum pump and bulk milk tank. In fact, the latter form of measurement is the easier, given careful design, and has the merit that it provides a measure of yield in the preferred recording units (kg). Volume measurement is less favoured because it is temperature dependent.

The incorporation of a strain-gauged beam in the jar's bottom support cradle, as shown in fig. 7.7(a), is one method of milk weighing which has proved reliable and capable of ± 0.05% accuracy, given that the arrangement of the jar's external vacuum and milk tubes minimises variable drag on the weighed jar.

In operation, a microprocessor circuit tares the weight of the empty jar before each cow is milked. When a push-button is pressed the following sequence begins. Valve 1 closes, freeing the clusters for attachment to the cow. Valve 2 closes and 3 opens. After two minutes the strain gauge output is sampled regularly, to monitor the milk flow. When this falls below 0.5 kg/min there is a delay, dependent on flow rate, after which the net strain gauge output (i.e. milk yield) is stored and the valve positions reversed, to release the clusters and the milk.

The milk level in the jar can be measured internally with a transducer in the form of a vertical, stainless steel rod, fitted with a sliding

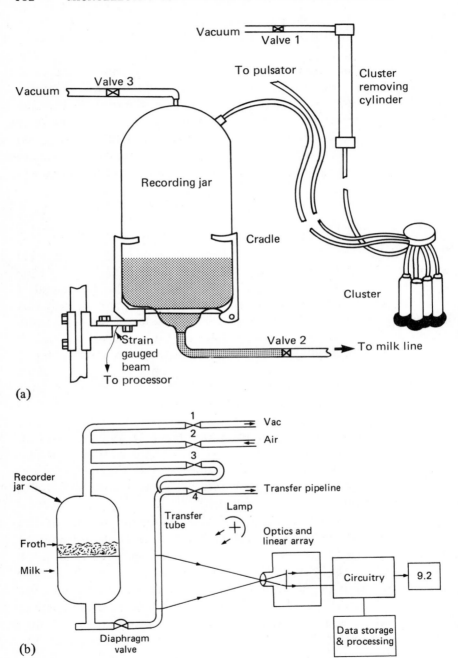

Fig. 7.7 Milk yield recording — recording jars. (a) Weight recording with strain-gauged support (b) Level recording, with photodiode array (NIAE, NIRD).

float. The position of the float is detected electronically and is recorded at the end of each cow's milking, before her milk is transferred to the bulk tank. This measurement is insensitive to foam at the top of the milk.

The static parlour, such as the popular herringbone parlour, requires an automatic yield meter on each recording jar, just as it needs a separate cow recognition unit and (if used) concentrate dispenser at each stall position. The rotary parlour has an advantage in this respect. Each cow can be identified by a single archway unit as she enters a stall to start on her circuit of the parlour. Her stall position is known from then on and a single high-rate concentrate dispenser at the entrance is all that is needed to deliver each preset ration, for the cow to eat while she moves round. The milking cluster is attached near the start, too, after the usual attention by the parlour operator to teat hygiene, and by the time that the cow reaches the exit position her milking should be complete.

A single milk metering device near the exit of the parlour is therefore sufficient, in principle. It must measure the yield in each jar as it passes by, before the milk is transferred to the bulk tank. This can be done gravimetrically by lifting and weighing the jar at this point or by transfer of the contents of the jar first to an intermediate weigher and then on to the tank. Alternatively, it is possible to return to the more traditional method, i.e. level measurement, replacing the operator's eye by a photoelectric device.

Figure 7.7(b) shows a photoelectric 'camera' which has been developed for milk level measurement in the rotary parlour. The problem of froth at the top of the milk is overcome by use of the side transfer tube. The bottom diaphragm valve is opened automatically by a sequential control circuit, about twenty seconds before the reading is taken. The open pinch valve 3 allows pressure to be equalised between the jar and the transfer tube (pinch valve 4 is closed). The level of milk in the transfer tube is therefore steady by the time that the camera views it. After the reading has been taken the control circuit closes valves 1 and 3 and opens 2 and 4, thereby initiating transfer to the bulk tank. The camera, which can be 750 mm away from the jar, contains a lens which focuses an image of the transfer tube on a linear, self-scanned array of 128 photodiodes, about 25 mm long overall. A photoswitch activates the measuring system as the jar reaches the measurement point, then the photodiode array is scanned by an electronic circuit, to produce an output signal with a duration dependent on the height of the milk in the tube. The duration of the signal is measured by causing it to 'gate' pulses from a clock generator into a counter. The clock frequency can be scaled to produce an output scaled in any units required, e.g. litres or kilograms. The accuracy of measurement is about ± 2%, which is well within the limits set by the U.K. Milk Marketing Board. The MMB requires that at any milking 95% of the measurements must not deviate more than ± 5% or 0.25 kg (whichever is greater) from the true weight.

7.5.2 Pipeline meters

Measurement of the irregular milk flow from the milking cluster on the cow is not easy. A pipeline flowmeter must not only cope with the air/milk mixture in the pipe but also conform to strict standards of performance. It must measure to the required accuracy (see 7.5.1) at flow rates up to about 7 kg/min; it must be easy to sterilise by the normal circulatory cleaning and sanitising equipment, in order to meet the standard of hygiene required in milk handling, and it must not impede milk flow by more than a stated amount. An ISO standard limits the pressure drop caused by the introduction of the meter to 3 kPa at a flow rate of 3 kg/min, for example.

These requirements are met by several different types of meter. In general, they operate by repeatedly collecting a small, fixed weight or volume of the milk in the pipeline, then releasing it towards the bulk tank, incrementing a counter on each occasion. The total count at the end of a cow's milking is then a measure of her yield. This counting method immediately lends itself to electronic systems of parlour monitoring and control. All that is needed is an electrical pulse output from the meter at each increment of weight or volume.

A dual tilting tray mechanism is employed in some meters – one tray filling as the other empties. Plate 20 shows a pipeline meter of a different type. A small chamber in the pipeline unit is repeatedly filled to one level and depleted to another, using conductivity probes (1.3.6) to define these two levels, which determine the volume of milk discharged at each cycle (0.2 l in this case). The associated meter unit displays the integrated flow in kilograms, holding the final reading until the cluster is attached to the next cow.

Plate 20 Milk yield recording – pipeline meters. (a) Pipeline unit (with milk sampling attachment) (b) Meter, with facilities for linking to a printer or computer. Approved by UK Milk Marketing Board for milk recording and butterfat sampling (Fullwood and Bland Ltd.).

7.5.3 Bulk milk meters

Flowmeters of a more conventional kind can be used when the contents of the farmer's bulk milk tank are transferred to the bulk tanker. The tanker pumps the milk at a high rate (at least 200 kg/min) and with a steady flow. For this application a turbine meter with electrical detection of blade rotation (1.3.3) is satisfactory from the hygiene standpoint but a special design has been evolved for milk metering because, inter alia, payment to the farmer and distributor is based on this measurement. A standard of ± 0.35% at the farm and ± 0.25% at the dairy or creamery has been set by the U.K. dairy industry. The standard U.K. meter contains an electronic counter and a ticket printer, which logs the amount of milk transferred at each end of the tanker's journey. The U.K. Milk Marketing Board has set up a comprehensive, computer-controlled facility for regular calibration of these meters.

7.5.4 *Milk sampling and analysis*

Periodically, each cow's milk is sampled for subsequent determination of its butterfat content. This requires that on the recording day an accurate and small proportion of the whole milk flow from every cow is collected at each milking. The amount gathered (about 50 g is recommended) is 1% or less of the total yield under normal circumstances and this must be representative of all the cow's milk, as already stated, or in the U.K. it will fail to meet the requirement that 95% of the samples must not deviate more than ± 0.1 percentage butterfat units from the 'true test'.

Sampling is achieved by precise partition of the pipeline milk flow, followed by collection of the sample in a jar below the flow divider. The remainder of the operation is performed manually and this seems likely to continue as long as the samples are analysed off-farm.

Milk analysis is still a job for the specialist laboratory but instruments which measure the absorption of infra-red radiation by homogenised samples of the milk simplify and speed up the analysis. These can measure milk fat, protein, S.N.F. (solids not fat), water and other constituents and up to four components can be determined in under half a minute. Wavelengths in the 3 μm to 10 μm range are employed. It will require the development of inexpensive emitters and detectors before this analysis will be economic on the farm.

7.5.5 *Automatic cluster removal*

Automatic cluster removal at the end of a cow's milking has been widely adopted as a labour-saving device, particularly valuable in rotary parlours. A flow sensor is employed to detect the end-point of the milking, and to actuate the vacuum-powered cluster remover (A.C.R.) via a mechanical or electrically-operated valve. End of milking is usually taken to be the point at which milk flow has diminished to about 0.2 l/min. This is not always a clearly determinable point − the flow may drop to 0.2 l/min and then increase again before returning to the threshold level. Therefore a time delay is built into the system, of sufficient length to avoid premature cluster removal but not long enough to risk over-milking, which is equally to be avoided.

Electronic sensing of milk flow by any of the methods described in 7.5.1 and 7.5.2 provides a signal which can be used for cluster removal. As already noted, the output of the strain gauged beam for automatic weighing of a recording jar is employed in this way, the delay being lengthened if the flow rate is not down to 0.2 kg/min in 40 s. The pipe-line meter in plate 20 causes the clusters to be removed if the 0.2 l cell does not fill in 45 s. Although mechanical systems for cluster removal are widely employed, electronic control provides greater flexibility in setting the end-of-milking threshold and in determining when the clusters should be removed.

7.6 Health monitoring

A cow's weight is one general indicator of her health, as is her appetite. Means of checking these factors have been described in 7.3.3 and 7.4. As stated in the introduction to this chapter, farmers are also concerned about their ability to detect the onset of oestrus and mastitis infection. Failure to recognise either of these very different conditions in good time is a cause of financial loss, apart from the ill-effects borne by the cow with a clinical case of mastitis. Many attempts have been made to detect these conditions automatically through electronic sensors, and some have produced encouraging results which are outlined in this section. Nevertheless, it must be acknowledged that biological scientists are not universally convinced by these measurements, on the grounds that other factors may influence the results to such an extent that they cannot be relied on. The dispute is unlikely to be resolved until further evidence has been acquired through more extensive trials.

7.6.1 Oestrus and pregnancy

Small-scale trials with temperature sensors built into the central claw-pieces of milking clusters have shown some correlation between milk temperature and the onset of oestrus. Average milk temperatures at morning and evening milkings were built up by a method similar to that used for cow weight averaging (7.4.2) and each new morning or evening reading was examined for significant departures from the appropriate average. In most cases an increase of about 0.3°C was found at oestrus. This resolution is well within the capability of high quality electrical temperature sensors (1.3.4). Therefore, if measurement of milk temperature provides a reliable guide to oestrus — either through the simple averaging above, or through more complex analysis — the electronic equipment is already available.

Laboratory tests on milk samples (for progesterone) now make it possible to detect non-pregnancy in a cow with over 98% accuracy at twenty-four days after insemination. This early warning of unsuccessful insemination is invaluable, since it allows the farmer to return the cows concerned to service with the minimum of delay. No rapid physical method has been developed for pregnancy detection but there are several pointers to possible methods which could be used on farms if suitably robust and inexpensive sensors were available. For example, absorption measurements on cervical mucus in the ultraviolet region, at 278 nm, revealed marked differences between oestrous, dioestrous and pregnant cows. However, in order to minimise false pregnancy diagnoses knowledge of the dioestrous level in each cow is necessary and there can still be a few misleading diagnoses as a result of abnormalities in some beasts. The analysis of the data on the farm would therefore call for microprocessor-based equipment with individual cow records in its memory.

7.6.2 *Mastitis detection and control*

Despite the use of antibiotic therapy and other control measures, mastitis continues to cause heavy financial loss through reduced milk production and replacement of infected cows. Severe clinical infection is evident when clots appear in the milk and these can be observed by the parlour operator with the aid of transparent pipeline units which contain metal mesh filters. If infection could be detected and treated at an earlier stage, however, new infections would be reduced, although at an additional cost in cow treatments and milk rejected. Therefore the search continues for a method of early detection which will respond to the changes brought about by the primary pathogens causing mastitis (staphylococci, streptococci, etc.) rather than the secondary pathogens which rarely cause clinical mastitis or produce reduced milk yields.

Milk monitoring   Temperature measurements in the clawpiece of the milking cluster have revealed sharp increases — up to 2°C — in milk from cows with udder infections, before the condition was evident to the eye. However, most attempts to produce an automatic mastitis detector have been based on measurement of the electrical conductivity of the milk. Two of the changes in milk which are brought about by mastitis are increases in the concentration of somatic cells and of the electrically charged ions which are normally present at a fairly constant concentration. In particular, the levels of sodium and chloride ion concentrations may increase by 50% to 100%.

Ionic changes may be monitored with ion-selective electrodes (4.3) if they can be developed to the required level of stable operation but until then the overall ion concentration can be determined conducti-metrically. It is essential to make this measurement separately in the pipeline from each teat cup to the clawpiece of the milking cluster, since the effect of mastitis is likely to appear in one quarter of the cow's udder before spreading to the other three. Without com-parisons of the four measurements variations due to other causes and affecting all quarters could mask the effect of mastitis. In fact, some research workers have gone further and made day-to-day com-parisons of conductivity in each quarter, to increase the reliability of detection and to reduce the number of false positives recorded. These analyses depend on computer processing of stored measure-ment data.

An additional factor is the difference in conductivity between the foremilk and strippings (the first and last milk from the cow at each milking, respectively). This difference, as well as the absolute con-ductivity, depends upon the interval between milkings. Nevertheless, measurements on strippings have correctly identified nearly 97% of infections in a trial herd, compared with about 82% with measure-ments on foremilk.

Conductivity measurement is relatively straightforward but its temperature dependence requires an auxiliary temperature sensor, to provide the necessary correction (1.3.6). Although stainless steel conductivity probes are used in direct contact with milk, without causing problems of hygiene (cf. 7.5.2), the measurement is some-times made at radio-frequency, instead of the usual audio-frequency, using non-contacting electrodes. These are fitted on the outside of a section of the flow tube (which must be of an electrically insulating material). The measurement cell thus formed is made part of an r.f. bridge circuit.

As in the case of oestrus detection, therefore, electronic equipment can collect and process information from which affected cows can be distinguished, although more confidence in the reliability of the diagnoses is needed. Meanwhile, electronics contributes to two of the control measures alluded to at the start of this sub-section. These are outlined next.

Automatic teat spraying   Regular udder washing with water or disinfectant before milking contributes to clean milk and the reduction of teat infections, if the cows' teats are dried manually or automatically directly after washing. Post-milking teat disinfection is equally important as a measure against intramammary infection but to be effective thorough coverage of each cow's teats is necessary, preferably with the formation of a drop of disinfectant at the teat ends after treatment. Hypochlorite solutions are suitable for this operation but create irritation to men and animals in confined spaces, therefore automatic post-milking spraying is best performed in a race at the parlour exit. Spray jets are mounted just above floor level, so that the cow has to straddle them as she walks through the race. This helps to expose the udder to the spray.

Some automatic teat spraying systems employ a sequence of photocells to detect the passage of the cow along the line of jets, switching on each jet in turn. This helps to give maximum coverage with economical use of disinfectant solutions (fig. 7.8).

Vacuum control   It is a widely held view that the severity of the vacuum fluctuations to which the cow's teat is exposed during milking may be a factor in the incidence of mastitis. The alternate suction and release of the teat cup liner which grips the teat is sometimes applied simultaneously to all four teats and sometimes alternately to pairs of teats. Alternate pulsation is advantageous,

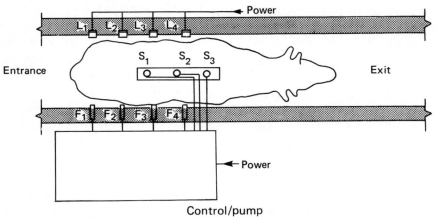

Fig. 7.8 Automatic walk-through teat disinfection system (L, lamps; F, photocells; S, spray nozzles) Sequence of operations: (i) All lamps obscured (as shown). Spray pump primed. (ii) L_2 to L_4 obscured. S_1 activated (1 s). (iii) L_3 and L_4 obscured. S_2 activated (1 s). (iv) L_4 obscured. S_3 activated (1 s). (v) No lamps obscured. Switch off. (Gascoigne Milking Equipment Ltd.).

because it results in a more stable vacuum. Vacuum demand can be reduced and its level stabilised with an electronic controller with three phased outputs, which control separate sets of pulsators (plate 21).

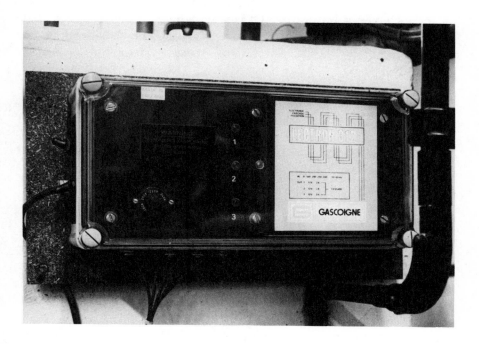

Plate 21 Cascade pulsation unit for milking machines (Gascoigne Milking Equipment Ltd.).

Further benefits are obtained by programming changes in the vacuum level, the pulsation rate and the pulsation ratio (the suction phase as a percentage of the total cycle) as milking proceeds. One commercial system employs a vacuum of about 35 kPa at 48 pulsations/min at the beginning and end of milking, rising to 50 kPa at 60 pulsations/min for the main period of milk let-down.

An experimental electronic controller has gone further in moderating the effects of vacuum fluctuations on the cow's teats. The pulsations are applied to each teat cup in turn rather than alternately or simultaneously, through a clock-driven, digital circuit, incorporating PROMs which store details of the required pulsation rates and ratios. The output from the controller operates electromagnetic relay pulsators or solenoid valves.

7.7 Monitoring and control systems

Sections 7.2, 7.3.3, 7.4.2, 7.5 and 7.6 have all dealt with elements of automatic monitoring and/or control in the dairy parlour, or closely associated with it. Some farmers are content to employ one or more of these elements, but not all. Others, with larger herds, are interested in comprehensive systems, sometimes linking the data from more than one dairy herd. Several of these systems are available commercially.

Figure 7.9 is a diagram of one commercial system, applied to a herringbone parlour, which could contain up to thirty-two milking points. It will be seen that the electronics equipment is divided between an office and the parlour (the latter, in fact, including the associated out-of-parlour sites). Automatic identification of each cow takes place when she enters one of the stalls. A coil fitted to the plastic manger energises the animal's transponder and an aerial (antenna) at the entry (right-hand) end of the parlour picks up the re-radiated code signal from each stall in sequence, each time a new set of cows is let into one side of the parlour or the other. A microprocessor-based unit decodes the information from the identification points in the parlour and at any external walk-through weigher or out-of-parlour feeder. Every identity is checked by the processor, so that erroneous identification is virtually eliminated. A second microprocessor unit processes the signals from the strain-gauge milk yield sensors, determining the point at which the clusters are removed automatically and measuring the tared weight of the recording jar as already described. Both sets of information pass to the third microprocessor in this distributed processing system, via a parlour control unit. The Farm Management System contains the third processor, which is also linked to the four peripherals shown.

The parlour control unit (shown enlarged) is linked to and controls the concentrate dispensers in or out of the parlour. Its associated keyboard and alphanumeric display units also allow exchange of 'notebook' information on individual cows between the parlour operator and the management system during milkings. The management system stores all the herd data onto triple floppy discs and further communications to and from the system are available through the office keyboard and VDU. A printer provides hard copy of the information required by the herd manager. A twenty-four-hour digital clock maintains the system in synchronism with the real world.

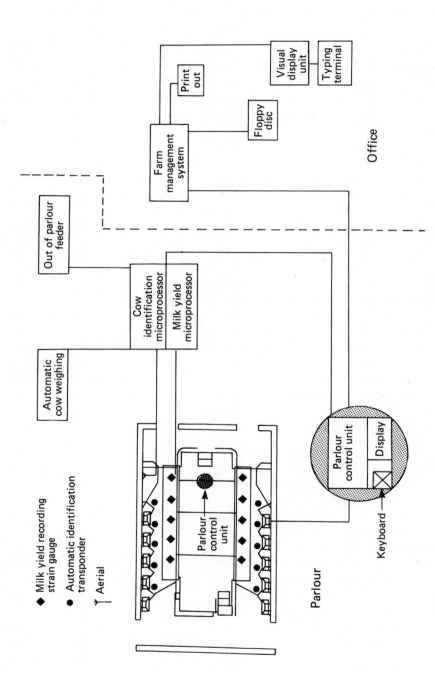

Fig. 7.9 Comprehensive parlour monitoring and control system, employing distributed microprocessors (Fullwood and Bland Ltd.).

The parlour operator is alerted to a message about an individual cow when she is identified on entering a stall. An audible warning sounds and a light appears at her stall position. The operator acknowledges, via a push-button, and the relevant message appears on the parlour control display. When the required action has been taken a second operation of the 'acknowledge' button cancels the display and the message. Some examples of the messages which can be exchanged are given below.

Note x is any decimal digit

R/L refer to the right and left halves of a herringbone parlour

Qr/Qrs refers to the affected quarter or quarters in the cow's udder

Message (to operator)	*Remarks*
1. Check for mastitis	Stall — R/L x; Qr/Qrs; Cow xxx
2. Treat for mastitis	Drug. This message will appear at the relevant milking for the course of treatment. Stall — R/L x; Qr/Qrs; Cow xxx
3. Dump infected milk	Mastitic (or other reason). This will continue for the required number of days after last treatment, in accordance with the drug being used, or veterinary advice. Stall — R/L x; Cow xxx
4. Dump colostrum	(Day 1, Day 2, Day 3, Day 4) Stall — R/L x; Cow xxx
5. Dry off	Treatment. Stall — R/L x; Cow xxx
6. Sample milk	Stall — R/L x; Cow xxx
7. Incorrect group	(Irrelevant identity) Stall — R/L x
8. Dry cow	Stall — R/L x; Cow xxx
9. Check yield	Stall — R/L x; Cow xxx
10. Feeder fault	Stall — R/L x
11. Strain gauge failure	Stall — R/L x
12. Identity not entered	Stall — R/L x
13. Check oestrus	Stall — R/L x; Cow xxx
14. General feeder fault	
15. Excessive feed set	Stall — R/L x; Cow xxx
16. Yield/identification system failure	

Message (by operator)	*Remarks*
1. Mastitis suspected	Noting the affected Qr/Qrs. This will then appear for the following two milkings

2. Mastitis confirmed   Noting the drug used
3. Oestrus cow
4. Cow requiring attention Lame, cut teat, off-colour
5. Milk dumped   Blood in milk
6. Draft vet   Cow xxx

(Details by permission of Fullwood and Bland Ltd.)

In addition, action lists and events lists can be called up on the VDU or printed. These can be daily, weekly or monthly and include lists of cows requiring or having received treatment; details of services and calving; milk sampling; total concentrates fed; total milk yields and feed conversion in kilograms of concentrates fed per litre of milk produced. The data bank may also include full details of all herd lactations, cow pedigrees and progeny of a given cow or bull. Financial and other business programmes can be added to the management system. The parlour therefore contains a very powerful management tool, as well as an aid to operator efficiency and herd health.

Power supply back-up is essential for this type of system, of course, and even so it must allow full manual control over any of its elements in the event of failure of the associated transducers or electronic circuits.

7.8 Waste treatment

Section 6.7 outlines the application of electronics to monitoring and control of biogas production from pig slurry. Cattle farms usually have more land than pig farms on which to spread slurry or farmyard manure (straw/slurry mixture) without causing offence to local residents or polluting watercourses. This has reduced the pressure on the cattle farmer to process the waste from his animals — except perhaps to separate the fibrous solids from the rest, in the interests of more efficient storage and better use of the material as a nutrient for grass and arable crops. Pig and cattle slurries have some physical, chemical and biological differences, too. Nevertheless, biogas can be generated from either in much the same way and some production plants are based on slurry from large cattle herds. The potential for electronic monitoring and control is the same in all cases.

7.9 Future developments

In looking ahead it is convenient to take the subjects of sections 7.2 to 7.8 in turn. Automatic animal identification will continue to hold the key to many of the likely developments in the cattle sector. The increasing market for digital coding devices capable of identifying millions of individuals uniquely is almost certain to lead to the production of tiny and inexpensive transponders or battery-powered devices which will be used to identify beef or dairy cattle and implanted for life, if required. The large bit-capacity of these devices will make it possible for each animal to carry its national herd number. The implanted transponder will identify the carcase of a beef animal, too. Farmers, breeders, veterinarians, processing and marketing organisations will all gain advantage from the ability to identify any animal precisely and reliably.

Feeding of bulk rations to a given dry weight, rather than total weight, should become an established procedure through the development of infra-red or microwave techniques. The further development of mass-produced sensors operating in the longer wavelengths of the infra-red region and, possibly, in the sub-millimetric waveband, may lead to more comprehensive on-farm determination of the quality of these rations. This would extend the capability of on-line monitoring and control systems.

There is reason to hope that experience with walk-through weighers outside the dairy parlour will lead to an increase in the percentage of valid weights obtained daily, through improvements in cow handling, coupled with improved data processing in the associated computer equipment. Beef cattle weighing should develop on these lines, too.

The expected range of new sensors for the longer infra-red wavelengths and beyond will have as much application in on-farm milk analysis as in bulk feed analysis. Milk monitoring in relation to cow health seems almost certain to develop, perhaps through on-line measurements in the ultra-violet region or through the use of ion-selective electrodes. The emergence of digitally-controlled milking machines seems equally likely.

The distributed computer network will also extend through the dairy farm to take in the additional information made accessible through new sensors, and to bring together all the management aspects of the enterprise, as well as increasing the applications of automatic control. Monitoring and control of biogas production could be one element in the comprehensive network.

7.10 Further reading

General
Castle, M.E. and Watkins, P. (1979). *Modern Milk Production*. London and Boston: Faber & Faber.
Dodd, F.H. *ed*. (1980). *BSAP Occasional Publication No. 2. The Mechanisation and Automation of Cattle Production*. Reading, Berks: The British Society of Animal Production.

Section 7.2
Bridle, J.E. (1976). 'Automatic dairy cow identification'. *Journal of Agricultural Engineering Research* 21, 41-48.
(1979). *Proceedings of 83rd Annual USAHA Meeting, San Diego, California*. Holm, D.M. and Araki, C.T. 'Status of electronic identification and temperature monitoring', 320-335.
Puckett, H.B., Hyde, G.M., Olver, E.F. and Harshbarger, K.E. (1973). 'An automated individual feeding system for dairy cows'. *Journal of Agricultural Engineering Research* 18, 301-307.
Street, M.J. (1979). 'A pulse-code modulation system for automatic animal identification'. *Journal of Agricultural Engineering Research* 24, 249-258.

Section 7.3
Broster, W.H. and Swan, H. (1979). *Feeding Strategy for the High Yielding Cow*. St Albans, Herts: Granada Publishing.
Bruce, D.M., Smith, R.A. and Meeks, I.R. (1980). 'Mobile feeding equipment — a weighing unit with auto-taring and batching facilities'. *Journal of Agricultural Engineering Research* 25, 155-160.
Dawson, J.R., Hooper, A.W. and Ambler, B. (1976). 'A continuous weigher for agricultural use'. *Journal of Agricultural Engineering Research* 21, 389-397.
Turner, M.J.B. (1980). *Report No. 34. Conveyor Feeding of Cattle*. Silsoe, Beds: National Institute of Agricultural Engineering.

Section 7.4
Filby, D.E., Turner, M.J.B. and Street, M.J. (1979). 'A walk-through weigher for dairy cows'. *Journal of Agricultural Engineering Research* 24, 67-78.

Section 7.5

Hoyle, J.B. (1965). 'The accuracy of milk meters'. *Journal of the Society of Dairy Technology* **18**(3), 171-175.

Sobel, A.T., Scott, N.R. and Hoffman, G.W. (1980). *ASAE Paper No. 80-5526 Automatic Milk Yield Data Acquisition*. St Joseph, Michigan: American Society of Agricultural Engineers.

Thiel, C.C. and Dodd, F.H. *eds*. (1977). *Machine Milking*. Shinfield, Berks: National Institute for Research in Dairying.

Section 7.6

Cant, E.J. and Reitsma, S.Y. (1979). 'A programmable pulsator control unit for milking systems'. *Journal of Agricultural Engineering Research* **24**, 331-336.

Fernando, R.S., Rindsig, R.B. and Spahr, S.L. (1980). 'Electrical conductivity of milk for detecting mastitis'. *Illinois Research* **22**(1), 12-13.

Heap, R.B. (1976). 'A pregnancy test in cows from progesterone in milk'. *Journal of the Royal Agricultural Society of England* **137**, 67-76.

Hyde, G.M., Puckett, H.B., Olver, E.F. and Harshbarger, K.E. (1981). 'A step towards dairy herd management by exception'. *Transactions of the American Society of Agricultural Engineers* **24**, 202-207.

Linzell, J.L. and Peaker, M. (1972). 'Day-to-day variations in milk composition in the goat and cow as a guide to the detection of sub-clinical mastitis'. *British Veterinary Journal* **128**, 284-294.

Maatje, K. and Rossing, W. (1976). 'Detecting oestrus by measuring milk temperatures of dairy cows during milking'. *Livestock Production Science* **3**, 85-89.

Marshall, R., Scott, N.R., Barta, M. and Foote, R.H. (1979). 'Electrical conductivity probes for detection of estrus in cattle'. *Transactions of the American Society of Agricultural Engineers* **22**, 1145-1151.

Scott Blair, G.W. (1956). 'Physical properties of cervical mucus in relation to bovine fertility'. *Netherlands Journal of Agricultural Science* **4**, 104-106.

Section 7.7

Puckett, H.B., Olver, E.F., Spahr, S.L. and Siktberg, T.J. (1980). *ASAE Paper No. 80-5022. Use of a Micro- and Mini-computer System for Real-time Data Collection*. St Joseph, Michigan: American Society of Agricultural Engineers.

8 Computers in Farm Management

8.1 Introduction

In earlier chapters repeated reference has been made to the use of computers as farm management aids. In most cases this has related to an on-line microcomputer, providing management information by processing data from monitoring instruments. However, off-line data processing of the kind now common in business offices is finding wider application in agriculture and horticulture, through the availability of ever more compact and powerful microprocessor-based business computers with diminishing hardware costs and a growing variety of standard software packages.

The development of data banks and data communications links are other factors which are certain to have a substantial influence on the type and range of applications of farm electronics equipment in the 1980s. In this sphere the farmer and grower can expect to benefit from equipment and services marketed for the wider industrial world, although some services will be specific to agriculture and horticulture.

This short chapter is primarily concerned with software, since most farmers and growers are likely to buy or hire computer packages which will run on standard business machines. However, it would be incomplete without a survey of discernible present trends in hardware and data networks. The two following sections deal with two not entirely separable types of software activity, namely processing and retrieval of information and forecasting based on mathematical models, respectively. The latter activity involves calculations of probabilities and risks on the basis of specified assumptions and for this reason is sometimes regarded with scepticism by farmers and growers only too aware of the vagaries of weather, markets and the influence of political decisions, nationally and internationally. Nevertheless, many farmers now acknowledge the value of operational research as an aid to planning of their enterprises. A third section deals with data communications and the chapter ends with an un-computerised forecast of future developments in the hardware and software spheres.

8.2 Data processing and retrieval

8.2.1 *The farm office computer*

Accounting and financial management by on-farm computer, as a separate activity, competes economically with a variety of secretarial and accountancy services, but is likely to show clear advantage on a farm with several substantial enterprises or on a group of farms under the same management. The computer's obvious advantage over the occasional up-dates of information provided by service agencies is the availability of the latest management information at any time. Timely application of this information can be turned to economic advantage.

Companies specialising in farm management computer systems will supply the minicomputer with a VDU, bulk storage (usually on disc) and a printer of the required graphics capability and print quality. In particular, the printer may produce standard pay slips which satisfy government regulations. It may also fill out cheques and remittance advice to suppliers, leaving only a signature to be added manually. Packages vary from one company to another, inevitably, but the customer can expect the computer to file manually entered information on creditors, debtors, income, expenditure, tax liabilities, forecast outgoings and returns etc. It should auto-matically up-date all relevant ledgers when a new entry is made. For example, information on a sales invoice will affect files on sales, debtors, gross margin, profit and loss, inter alia. Reference to the identifying number of the source document (the invoice, in this case) will be logged. Daily transactions may be listed and printed. Monthly financial statements may be in summary or detailed form. Yearly (and on demand) statements of profit and loss will include gross margins, fixed and variable costs and comparisons with the budget and performance in the previous year. Considerable flexibility, combined with reliability, is incorporated in these packages.

In addition, the computers may be used for stock control (seeds, plants, fertilisers, spray chemicals, livestock feed) and valuation of machinery and fixed assets, as well as off-line enterprise monitoring. The last-mentioned is particularly valuable to managers of livestock enterprises and growers of horticultural crops. The livestock farmer can maintain records of animal breeding (including services, pregnancy tests, abortions, births and pedigree lines), animal performance (including feed conversion, milk and egg yields, litter sizes and weights) and animal health (disease, injury and treatment). In addition, given the necessary analysis of his available feed, he can calculate the nutrient requirements of groups of animals of particular age and/or weight in order to determine their appropriate ration. With the aid of a linear programme (8.3) he can go further and compute a least-cost ration. The horticultural enterprise provides opportunities for planning the raising of plants and the production of cuttings in addition to recording inputs and yields.

The suppliers of off-line computer systems provide training sessions for farm staff, on the farm, as part of the implementation of any software package. The trend is to increasingly 'user friendly' software, which offers 'menus' and simple yes/no choices to the lay operator at each stage of a programme, thereby stimulating confidence in the beginner. Competence in handling the system is normally achieved in a few days. Carefully prepared manuals are very important in this context.

These systems may be bought or leased. The buyer can obtain a hardware maintenance contract and either type of user can pay for major software up-dates as they arise. The cost of these services is substantial, however − it may be as much as the initial capital cost over a five-year write-off period.

Although the off-line computer system was originally developed as a farm office machine its use for enterprise monitoring has brought it close to the automatic monitoring equipment described in earlier chapters. In fact, some have the necessary hardware to accept inputs from automatic data collection devices and require only minor modifications to their normal software to permit both manual and automatic data entry. Therefore, given a major input from automatic monitoring equipment the farm office computer becomes the host in a distributed processing system, maintaining the farm's own data bank. All it lacks at this stage is the ability to initiate on-line control action, which it must leave to local processors at the monitoring points. This ability to collect and store data from a distributed network enhances the economic justification for the office minicomputer considerably.

To end this sub-section reference must be made to semi-automatic data collection. Hand-held, battery-powered data loggers with manual keyboards are a very useful aid to farm staff who need to gather substantial amounts of information routinely in the field or in crop or livestock buildings. These units can be programmed for particular tasks, to ensure that they assemble the data in the required order for subsequent processing. Processing of the stored data by computer can be done via a telephone link, using an acoustic coupler attached to the telephone handset.

8.2.2 *Data banks and data bases*

Widespread access to data banks of many kinds is available from video/teletext services, using the home TV set or a VDU terminal to receive information broadcast or via the telephone, respectively. Current market and weather information can be conveyed in this way. Nevertheless, the user must either accept the selection of information broadcast or choose from a set of 'pages' of information in the case of the telephone system. Those familiar with information retrieval more generally will be aware of the value of data base searches via keywords which can restrict a search to one quite specific topic. National and international data bases contain a vast number of references to published information, often with informative abstracts. They can also refer the user to banks of scientific and engineering data. The coverage increases year by year.

Although special terminals can be used to interrogate data bases via the international telephone network this activity provides another potential application for the farm office computer. Set up in this way the computer will be able to search for information on many kinds of plants, soils, animals, diseases, weather patterns, engineering designs and so on. Data base searching invokes charges for connection time, therefore the operator seeking information on a specific topic must look at the keywords and work out a search strategy beforehand. Efficient searching comes with practice but initial training is available from the data base companies (appendix 3).

8.3 Computer prediction models

Operational research (OR) techniques, including linear programming, have been employed for many years as a management aid to decision-making. The OR specialist sets up mathematical models with which to predict the effects of possible changes in an enterprise of one kind or another. The model can be used with data from records of events and/or with notional data, the latter providing a means to predict the outcome of an expected or proposed development.

In the agricultural sphere major agrochemical companies and animal feed compounders have well-established linear programme services to farmers and in the U.K. the advisory service of the Ministry of Agriculture, Fisheries and Food also uses OR techniques. A few examples of OR models should suffice to show their range of applications.

Arable farm linear programme   One extensive model allows the user to specify the crops that he wishes to consider, together with the gross margin, the job sequence and the possible rotations for each crop. He must also specify the farm area and — for each job — the work rate, men and machines needed (with their costs), the periods when it can be done and the penalty for lack of timeliness, all according to his best estimates. The model examines the use of men and machinery in successive periods of the year, the sequencing of crops and of crop rotation, taking into account the economic value of timeliness and crop rotation. It then calculates the combination of cropping, men and machinery which gives the most profit.

Annual cost of machinery   If the user specifies the type and costs of each of his machines, their annual use, years to be kept, interest rate and estimated rate of inflation the model will calculate the annual cost of each machine. In this way the farmer can compare the estimated annual increase in profit from the introduction of a new machine with its annual cost.

Costs of cereal harvesting   A simulation model of cereal harvesting totals the operating and capital costs of a combine harvester, wet-grain store and drier and takes into account grain shedding, combine header and threshing losses. The model also simulates ten harvest years using recorded weather data from the appropriate area, to determine when harvesting is possible. Beginning with specifications of the day on which harvest starts, on the initial grain moisture content and on the farm's cereal growing area the model compares the cost per hectare of systems with differing drier and combine sizes, predicting which will give the minimum total cost.

Costs of dairy farming   The economic consequences of many changes in a dairy farm can be predicted in another large model.

For example, the farmer may wish to know the effect of an increase in the quality of winter feed from better silage, produced — say — by greater use of fertiliser and improved silage-making technique. Given the expected increase in feed value, in MJ/kg, the model can be used to calculate the herd size that the same area of grass will support and the saving in concentrates and labour.

Broiler nutrition and environment   A broiler growth model is the basis on which the feeding policy and environmental regime for optimum profit can be predicted. An equivalent model is available for egg production.

The application of any of these models to particular circumstances involves complex calculations, often with many repetitions. Here programmable calculators and computers of various sizes play their part. The programmable calculator, with a set of PROM and EPROM packages, meets the needs of simpler OR models. The farm office computer can cope with both small and medium size models but at present the largest programmes have to be run on main-frame machines. However, the development of computer networks makes it possible for the farm office computer to function as a remote terminal to a main-frame machine. In principle, therefore, a farmer or grower can gain access to powerful OR routines.

8.4 Data communications

Communication of digital data, with minimal corruption of the information content, is a subject of immense importance to governments, industry and science alike. Massive investment goes into communications developments and the technology is developing very rapidly. Long distance transmission via satellites, at very high frequencies, and via optical fibres, at even higher frequencies, make it possible to convey vast amounts of multiplexed information on single links. Semiconductor technology is playing its part in these developments — for instance, high intensity light-emitting diodes, capable of switching at high speed, are becoming available for fibre optics links. At the more traditional level, rates of data transmission via the telephone/telegraph network varies from the Telex 50 bit/s, through 2400 bit/s and above on the public switched network, to 1M bit/s or more on dedicated digital lines. The maximum rate depends in part on the use of a switched or dedicated line (the latter is always superior) and on the type of modems (modulator/

demodulator units) which couple the data transmitter to the line at one end and the line to the receiver at the other.

The agricultural and horticultural user of transmitted data needs to be aware of these services in a general way, in order to take best advantage of them. In this connection, serial data transmission between computers and peripherals, and in computer networks, usually conforms to the seven-bit (plus even parity bit) International Data Code of the International Organisation for Standardisation (ISO) − cf. appendix 2. This is derived from the American Standard Code for Information Exchange (ASCII). The voltages to be used in data transmission are standardised too and although these have been specified by the International Telephone and Telegraph Consultative Committee (CCITT) the equivalent U.S. Standard (RS232C) is most commonly quoted.

Serial data transmission brings with it the problem of synchronisation between the transmitter and the receiver, to ensure that the latter can recognise the beginning and end of each transmitted element. Although synchronisation of two mutually remote stations is possible much serial information is transmitted asynchronously but with a coded beginning and end of each element. Thus, all serial ISO seven-bit data words are framed by a start bit (high level) and one or two stop bits (low level). The bit-pattern of fig. 7.2(b) is in standard form.

Parallel data transmission between computers and peripherals has its standards, too. The American IEEE488 'bus' interface is pre-eminent in this sphere. Parallel transmission of binary words is faster than the serial version, of course, but more costly because it needs a multi-wire link. It is therefore used only for transmission over limited ranges, in general.

8.5 Future developments

The merging of off-line farm office computing with automatic monitoring and control on the one hand and external sources of information on the other seems inevitable. Regularly up-dated meteorological charts, information on the agricultural or horticultural enterprise itself and on the relevant market situation will all be to hand when it is needed. User-friendly OR packages will be available for the farmer or grower to test planning strategies and central data banks will provide much of the information needed during the planning phase.

The development of electronics and communications technology — high density disc or bubble memories, flat tube VDUs, fibre optics communication links, etc., will make many more of these activities economically possible on-farm in the future.

8.6 Further reading

Audsley, E. (1981). 'An arable farm model to evaluate the commercial viability of new machines or techniques'. *Journal of Agricultural Engineering Research* **26**, 135-149.

Audsley, E. and Boyce, D.S. (1974). 'A method of minimising the costs of harvesting and high temperature grain drying'. *Journal of Agricultural Engineering Research* **19**, 173-188.

Audsley, E., Dumont, S. and Boyce, D.S. (1978). 'An economic comparison of methods of cultivating and planting cereals, sugar beet and potatoes and their interaction with harvesting, timeliness and available labour by linear programming'. *Journal of Agricultural Engineering Research* **23**, 283-300.

Audsley, E., Gibbon, J.M., Cottrell, S. and Boyce, D.S. (1976). 'An economic comparison of methods of storing and handling forage for dairy cows on a farm and national basis'. *Journal of Agricultural Engineering Research* **21**, 371-388.

Audsley, E. and Wheeler, J.A. (1978). 'The annual cost of machinery using actual cash flows'. *Journal of Agricultural Engineering Research* **23**, 189-201.

Blackie, M.J. and Dent, J.B. *eds*. (1979). *Information Systems for Agriculture*. London: Applied Science Publishers.

Hebditch, D.L. (1975). *Data Communications. An Introductory Guide*. London: Paul Elek (Scientific Books) Ltd.

United Kingdom Post Office (1975). *Handbook of Data Communications*. Manchester: NCC Publications.

Appendix 1 Units and Standards

A1.1 The International System of Units (SI)

The full list of SI units, their symbols and their inter-relationships are given in 'Le système International d'Unités', published by the International Bureau of Weights and Measures in Paris (1970). An English version was prepared jointly by the National Physical Laboratory, U.K. and the National Bureau of Standards, U.S.A. This version ('SI. The international system of units') is available in the U.K. from HMSO. The document also lists other units which can be used with the International System although not strictly part of it. Preferred multiples and sub-multiples of units are included. The practical use of the system is explained in ISO 1000 (1973), 'SI units and recommendations for use of their multiples and certain other units', available in the U.K. as BS 5555:1976.

Those SI and other allowed units which have been used in this book (excluding one historical reference to feet and inches) are given below.

	Quantity	Name of unit	Symbol
SI base units	length	metre	m
	mass	kilogram	kg
	time	second	s
	electric current	ampere	A
	thermodynamic temperature	kelvin	K
SI derived units	frequency	hertz	Hz
	force	newton	N
	pressure	pascal	Pa
	energy/quantity of heat	joule	J
	power/radiant flux	watt	W
	electric potential	volt	V
	electric resistance	ohm	Ω

	Quantity	Name of Unit	Symbol
	conductance	siemens	S
	Celsius temperature	degree Celsius	°C
		(K − 273 approx.)	(1°C = 1 K)
SI supplementary	plane angle	radian	rad
units and units	time	minute	min
used with the		hour	h
SI system		day	d
	angle	degree	°
	volume	litre ($10^{-3}m^3$)	l
	mass	tonne (10^3 kg)	t
Units to be used	area	hectare (10^4m^2)	ha
temporarily with	pressure	bar (10^5Pa)	bar
the SI system			

The British Standards Institution's publication PD 5686:1978, 'The use of SI units', also allows for use in special fields, r/min (rotational frequency) and kWh (consumption of electrical energy), 1 kWh = 3.6 × 10^6 J.

The preferred multiples and sub-multiples of units used for the book are:

Factor	Prefix	Symbol
10^9	giga	G
10^6	mega	M
10^3	kilo	k*
10^{-3}	milli	m
10^{-6}	micro	μ
10^{-9}	nano	n

Note Regrettably, the electronics world is prone to use K (kelvin) for k (kilo)

A1.2 Other scales and ratios used

dB   The decibel scale is based on a logarithmic ratio, using the common logarithm (to the base 10). The relationship between two voltage signal levels V_1 and V_2 is frequently expressed on this scale, viz.

$$dB = 20 \log (V_2/V_1)$$

The scale is employed in electronics as a means to quantify frequency response (cf. fig. 1.1), and amplification or attenuation of signals, inter alia.

m.c.w.b.   The moisture content of a solid, expressed on the wet basis, is defined as

$$\text{m.c.w.b.} = m_W/M_T \times 100\%$$

where m_W is the mass of water in a particular amount of material and M_T is the total mass of the material, including its water content.

The precise definition of the water content of a material is not easy, however, since normally some of the included water is chemically bound to other constituents. This is not usually regarded as part of the moisture content. Most moisture meters rely on calibration against one of the standard methods of determination (cf. 3.4).

pH   Acidity and alkalinity of solutions, in units from 1 to 14. The pH of an aqueous solution is defined as the negative common logarithm of its hydrogen ion activity. The neutral point between acidity and alkalinity lies at about pH 7, the exact value depending on temperature. Solutions are acid below the neutral pH level and alkaline above it.

p.p.m. (parts/   The concentration of a substance in a liquid, per
million)   million units of mass or volume.

R.H.   Relative humidity is usually defined by the expression

$$\text{R.H.} = m/M \times 100\%$$

where m is the mass of water vapour present in a particular volume of air and M is the maximum amount that the same volume of air could hold at the same temperature (i.e. at saturation). On the assumption that the mass of the water vapour present in the air is proportional to its vapour pressure p, the expression can be rewritten

$$\text{R.H.} = p/P \times 100\%$$

where P is the saturation vapour pressure (S.V.P.) of the water at the same temperature.

When air is cooled to its saturation point, the temperature at which this happens is the dew point (unless it is below 0°C, in which case it is the hoar frost point).

v.p.m. (volumes/ million)   The fraction of the total volume taken up by one gas in a gaseous mixture, per million units of volume.

A1.3 The electromagnetic spectrum

The velocity of electromagnetic radiation in free space is approximately 3×10^8 m/s. It relates to the frequency, f, and wavelength of radiation, λ, in the electromagnetic spectrum in the following way:

$$3 \times 10^8 \text{ (m/s)} = f(\text{Hz}) \times \lambda \text{ (m)}$$

Frequency is the quantity usually quoted in the radio and TV wavebands but wavelength is almost universally adopted beyond the microwave region (i.e. towards the shorter wavelengths). The frequency and/or wavelength bands of the spectrum are summarised in the table below, in ascending wavelengths.

Spectral region	Approx. frequency range	Approx. waveband
gamma/X-rays		fm (very hard) to 10 nm (soft)
ultraviolet		10 nm to 400 nm
visible		400 nm to 700 nm
infra-red		700 nm to μm region
microwaves/TV/ radio	THz to kHz	sub mm to km

Appendix 2 Communications Standards

A2.1 ISO seven-bit code

The International Organisation for Standardisation first produced ISO 646-1973(E) '7-bit character set for information processing interchange' in 1973. This divides a 7-bit alphanumeric character into two sub-groups, viz.

$$b_7b_6b_5 \quad b_4b_3b_2b_1$$

where b_7 is the most significant bit. It then allocates characters to the 128 possible 7-bit groups in the following way.

$b_7b_6b_5 =$	000	001	010	011	100	101	110	111
$b_4b_3b_2b_1$								
0000	NUL	DLE	SP	0	@	P	`	p
0001	SOH	DC1	!	1	A	Q	a	q
0010	STX	DC2	"	2	B	R	b	r
0011	ETX	DC3	£	3	C	S	c	s
0100	EOT	DC4	$	4	D	T	d	t
0101	ENQ	NAK	%	5	E	U	e	u
0110	ACK	SYN	&	6	F	V	f	v
0111	BEL	ETB	'	7	G	W	g	w
1000	BS	CAN	(	8	H	X	h	x
1001	HT	EM	)	9	I	Y	i	y
1010	LF	SUB	*	:	J	Z	j	z
1011	VT	ESC	+	;	K		k	
1100	FF	FS	,	<	L		l	
1101	CR	GS	–	=	M		m	
1110	SO	RS	.	>	N	Λ	n	–
1111	SI	US	/	?	O	–	o	DEL

The $b_7b_6b_5$ sub-group covers the range 0 to 7 in equivalent decimal numbers and the $b_4b_3b_2b_1$ sub-group covers the range 0 to 15. ISO

646 identifies the position of each character in the above table by its decimal combination. For example, 5/2 (binary 101 0010) refers to R and 0/11 (binary 000 1011) to VT (vertical tabulation).

The meaning of the control characters ACK to VT is given below.

Location	Abbreviation	Meaning
0/6	ACK	Acknowledge
0/7	BEL	Bell
0/8	BS	Backspace
1/8	CAN	Cancel
0/13	CR	Carriage Return
1/1-4	DC	Device Controls
7/15	DEL	Delete
1/0	DLE	Data Link Escape
1/9	EM	End of Medium
0/5	ENQ	Enquiry
0/4	EOT	End of Transmission
1/11	ESC	Escape
1/7	ETB	End of Transmission Block
0/3	ETX	End of Text
0/8-13	FE	Format Effectors
0/12	FF	Form Feed
1/12	FS	File Separator
1/13	GS	Group Separator
0/9	HT	Horizontal Tabulation
1/12-15	IS	Information Separators
0/10	LF	Line Feed
1/5	NAK	Negative Acknowledge
0/0	NUL	Null
1/14	RS	Record Separator
0/15	SI	Shift-In
0/14	SO	Shift-Out
0/1	SOH	Start of Heading
2/0	SP	Space
0/2	STX	Start of Text
1/10	SUB	Substitute Character
1/6	SYN	Synchronous Idle
0/1-6, 1/5-7	TC	Transmission Controls
1/15	US	Unit Separator
0/11	VT	Vertical Tabulation

Twelve of the characters can be varied to suit national use and the preceding code table gives the U.K. version, with £ at 2/3 and $ at 2/4.

A2.2 Serial binary communication

Asynchronous transmission of binary information requires some means by which the (unsynchronised) receiver can recognise the beginning of each transmitted character. This is achieved by 'framing' each character with 'start' and 'stop' bits. The signal line goes to its high voltage state ('1') at the end of each frame and the receiver recognises the arrival of the next character from the start bit ('0') which precedes it. The receiver must be programmed to know the transmission rate in bit/s: it can then decode the 7-bit pattern and parity digit which follows the 'start' bit, by reference to its own clock. One or two 'stop' bits ('1's) return conditions to the start of the cycle in readiness for the arrival of the next character.

The interface between digital equipment (including keyboards, printers and VDUs) and two or four wire communications links has been standardised by the International Consultative Committee for Telephone and Telegraph (CCITT), which is part of the International Telecommunications Union. CCITT Recommendation V24 specifies the control, timing and data circuits which are needed to effect data exchange. Commonly, manufacturers work to an almost identical American standard, EIA RS 232C, 'Data terminal equipment and data communications equipment employing serial binary data inter-faces'. This was produced by the Electronic Industries Association (EIA) in 1969.

A2.3 'Bus' interfaces

Over shorter distances binary information can be carried on parallel signal lines with the benefit that it is transferred character by character, rather than bit by bit. The communications 'bus' regulates exchange of data in this form. Several forms of bus are in use but the American IEEE standard 488-1975 is the most widely recognised in engineering applications. An IEEE bus comprises sixteen lines, of which eight carry the data. The remainder are concerned with the control of data exchange and addressing of a particular receiver in a system with more than one line of communication.

Appendix 3 Advice and Training (U.K.)

A3.1 Microelectronics and microprocessors

General awareness courses on microelectronics for non-specialists — with or without an engineering background — are available at educational establishments in most regions of the U.K. Some also offer training for operatives who are required to work with microprocessor-based instrumentation, control and data processing equipment.

Courses specific to agricultural applications include some in the Short Course programme provided by the National College of Agricultural Engineering, Silsoe, Beds, MK45 4DT. These provide instruction and hands-on experience for the non-specialist.

The Agricultural Development and Advisory Service of the Ministry of Agriculture, Fisheries and Food (ADAS) provides an advisory service in several ways. Farmers and growers, in particular, can obtain this advice through the ADAS Liaison Unit at the National Institute of Agricultural Engineering, Wrest Park, Silsoe, Beds, MK45 4HS, or through the nearest Mechanisation Advisory Officer. The NIAE itself, with its sister-Institute, the Scottish Institute of Agricultural Engineering, Bush Estate, Penicuik, Midlothian, EH26 0PH, also provide information on this topic, inter alia, through the Association of the British Society for Research in Agricultural Engineering. The BSRAE is responsible for both NIAE and SIAE and the membership scheme is aimed at all sectors of agriculture and agricultural engineering.

Since the late 1970s manufacturers interested in applying microelectronics equipment to agricultural and horticultural operations have had the opportunity to apply to the Department of Industry for financial support under the Department's MAP (microprocessor applications) scheme. The DoI has also assisted NIAE in setting up a Product Evaluation and Testing Scheme, under which manufacturers can commission confidential evaluation of their equipment.

Information on these and other confidential services can be obtained from the Product Evaluation Group, NIAE.

A3.2 Measurement

Advice on sensors and electronic measurements, training in these subjects and confidential evaluation of measuring equipment can be obtained in the ways outlined in A3.1. In addition, advisory work is taken on by the Scientific Instruments Research Association (SIRA), South Hill, Chislehurst, Kent, BR7 5EH. From the agricultural and horticultural standpoint, SIRA has long experience in moisture determination, in particular. Calibration of instruments is undertaken by laboratories registered by the British Calibrations Service. The list of authorised laboratories can be obtained from BCS at the National Physical Laboratory, Teddington, TW11 0LW.

A3.3 Information services

Automation and instrumentation are two of the topics covered by the Agricultural Engineering Abstracts of the Commonwealth Agricultural Bureaux (CAB), Farnham Royal, Slough, SL2 3BN. These abstracts, prepared monthly at NIAE, provide a world coverage of major scientific and technical literature relevant to agricultural and horticultural engineering. The Information Services Group at NIAE can also be commissioned to carry out international data base searches from its computer terminal.

Those wishing to make direct use of an international data base such as the Lockheed Corporation's DIALOG, which provides access to a wide range of scientific and engineering abstracts, can do so via the telephone or a computer terminal. Training in data base searching is provided. In the case of DIALOG, this can be arranged through Dialog Information Retrieval Service, PO Box 8, Abingdon, Oxford, OX13 6EG.

Glossary

Analogue A representation of a variable by a physical quantity (e.g. a voltage) which is made proportional to the variable.

Analogue to digital (A/D) converter A device for converting an analogue signal (normally a voltage) to an equivalent digital representation of the signal.

BASIC Beginners All-purpose Symbolic Instruction Code. This is a high level language which is easy to learn and is generally used for personal computer systems. It uses English words as instructions and symbolic names for data.

Bit Binary digit. The basic data unit in digital data processing and computing. A bit may have the values 0 or 1.

Bus Group of wires, common to various units in a digital system, and used to carry data between them.

Byte An 8-bit binary data word.

Capacitance A measure of the electrical charge stored in an assembly of two electrodes, separated by a dielectric medium, when a potential difference exists between the electrodes. Numerically, the ratio of the charge (in coulombs) to the potential difference (in volts). Units : sub-multiples of the farad (F).

Capacitor An assembly of one or more pairs of electrodes, separated by a dielectric medium.

Central processor unit (CPU) The part of a computer which performs the arithmetic and logical functions. Has access to the programme being executed and the computer store.

Character One of a set of alphanumeric symbols which may be represented by a unique binary code pattern.

Chip Popular name for an integrated circuit device.

Clock   A regular train of pulses used to provide the timing control in a computer system.

CMOS   Complementary Metal-Oxide Semiconductor. A form of construction which produces electronic devices with low power requirement.

Common mode interference   Unwanted electrical signals induced from surrounding radiating sources (e.g. mains supply cables) equally into both conductors of a two-wire signal conducting system.

Common mode rejection (CMR)   The ability of a module or system to prevent a loss of accuracy by the presence of common-mode interference applied simultaneously with the signal. Defined as the ratio of the magnitude of common-mode interference to the magnitude of the signal.

Compiler   A computer program used to translate an application program written in a high level language, such as FORTRAN, into machine code form.

Conductance   With direct current, the reciprocal of the electrical resistance of a material. With alternating current, the resistance divided by the square of the impedance.

Data acquisition   The collection of data in any form (written or on paper tape, cards, etc.) from experiments, surveys or processes.

Data logging   The automatic or semi-automatic collection of data in printed or computer-compatible form. The term is normally restricted to systems where the data are recorded in digital form, thus excluding analogue chart or tape recorders.

Data processing   A systematic sequence of operations performed on data (e.g. merging, sorting, computing, manipulation of files) with the object of extracting information.

Debugging   The process of detecting and correcting errors in operation of a program.

Dielectric constant (permittivity)   A measure of the ability of a dielectric material to store electrical charge. Numerically, the factor by which the capacitance of a capacitor increases when the material replaces air as the dielectric medium.

Digital to analogue (D/A) converter   A device which produces an analogue representation of data when supplied with the data in digital form.

Diode   A two-electrode device which conducts electricity in one direction, by convention from *cathode* to *anode*.

Disc (disk) store   A magnetically coated disc used for mass storage of data in a computer system. Discs may be of the hard type, such as the Winchester disc, or may be of flexible plastic in a protective envelope as in the case of the floppy disc.

EAROM   Electronically Alterable Read Only Memory. A memory device in which the data can be altered electrically but will be retained in the memory when power is removed.

EEPROM   Electrically Erasable Programmable Read Only Memory. A read only memory that can be erased by applying an electrical signal to it.

EPROM   Erasable Programmable Read Only Memory. A read only memory that can be erased by using ultra-violet light or by applying an electrical signal.

File   A collection of related data stored in memory. Files may be held on floppy disc or on magnetic tape.

Floppy disc (disk)   A magnetically coated plastic disc mounted in a protective envelope and used to provide mass storage of data.

FORTRAN   FORmula TRANslator. A high level computer language intended for use in scientific applications.

Frequency modulation (f.m.)   An electrical system in which the frequency of an alternating signal varies in proportion to the value of the variable which it represents.

Frequency to voltage (F/V) converter   A device for converting a cyclic electrical signal into a voltage proportional to the cyclic frequency.

Hard disc (disk)   A rigid magnetically coated disc for mass data storage, such as the Winchester disc.

Hardware   The electronic and mechanical equipment that makes up a computer system.

High level language A computer programming language which uses English words as instructions and symbolic names for variables. Programs written in this language are translated into machine code by a compiler. Typical high level languages are BASIC, FORTRAN and PASCAL.

Hybrid circuit An electronic circuit built partly from integrated circuits and partly from individual components.

Impedance Total effective resistance of an electric circuit to alternating current. A combination of ohmic resistance, capacitance and inductance.

Impedance matching Matching the output impedance of one stage of an electrical system to the input of the next, in order to reduce power loss and distortion of transmitted signals.

Inductance A measure of the property of a circuit, or circuit component, to create a magnetic field and to store magnetic energy, when it carries a current. Unit — the henry (H).

Instruction A code or expression which defines what the CPU is to do and which data it shall use during the execution of the instruction. Instructions are made up of one or more bytes of data which are interpreted by the CPU prior to execution.

Integrated circuit (IC) A circuit in which all the components are formed upon a single chip of semiconductor material.

Interface The common boundary between units of a system, across which data can be transferred. The interface is required to make units compatible with respect to codes, operating speeds, signal levels, etc.

Kilobyte Term used for 1024 bytes. Thus a 4k byte memory will contain 4096 bytes. The kilo prefix may also be used in the same sense with bits and words.

LCD Liquid crystal display.

LED Light emitting diode.

LSI Large scale integration. Method of fabrication of integrated circuits which places a large amount of logic on to a single silicon chip. All microprocessors use LSI.

Memory Data storage hardware associated with the CPU. The memory may be RAM or ROM and may be used to store the program, or data being processed.

Microcomputer Term used to describe a complete computer system on a chip comprising the CPU, program memory, data memory and possibly several input-output ports. Also used for a complete micro-computer system made up from separate components.

Microprocessor The central processor (CPU) section of a micro-computer system.

Modem Modulator-demodulator. Device used to convert logic level signals into two audio tones for transmission over long distances and to restore the logic levels at the receiving end.

Multiplexing Process by which several different signals may be switched in sequence over a common set of wires.

Noise (electrical) Undesirable signals (interference) which distort or mask desirable signals.

Non-volatile memory Memory system which does not lose its data contents if the power is removed. Examples are the various forms of ROM and the magnetic bubble memory.

Off-line working Processing of data relating to a system with equipment not connected to that system.

On-line working Automatic collection and processing of data from a system, in real time.

Operating system The set of programs, usually stored on a disc, which govern the operations of the computer and includes assemblers, loaders, input-output utilities and file handling facilities.

Parity A method of error checking in which an extra bit is added to the data element. For odd parity the bit is set so that the total number of '1' bits in the data word is odd. It is also possible to use even parity.

Peripheral A device external to the processor, such as a printer, floppy disc unit or visual display unit which is controlled by and communicates with the CPU.

Program(me) The sequence of instructions to be followed by the CPU in order to carry out the desired operations.

PROM   Programmable Read Only Memory. Any read only memory that can be programmed in the field but the term is normally reserved for the fusible link types.

Pulse code modulation (PCM)   A method of transmitting data by repeatedly sampling a continuous signal, encoding each sample into a pulse train (usually in bit form) and transmitting the pulses directly or imposed on a high frequency (carrier) signal. This system is relatively immune to corruption by electrical noise.

RAM   Random Access Memory. Term used to describe read/write memory which is normally used for data storage in a processor system.

REAL TIME   A system operates in real time when data generated by an event are processed virtually simultaneously.

ROM   Read Only Memory. A memory which has a permanent data pattern written into it. ROMs are programmed by the use of links in the mask used when fabricating the chip.

Semiconductor   A material whose electrical conductivity is inter-mediate between that of metals and that of insulators.

Serial   Mode of data handling in which individual bits of the data word are dealt with in succession rather than simultaneously.

Series mode interference   Unwanted electrical interference induced into transducers and conductors and appearing at the measuring instrument in series with the signal.

Series mode rejection (SMR)   The ability of a module or system to prevent loss of accuracy by the presence of series mode inter-ference. Defined as the ratio of the magnitude of the series mode interference to the magnitude of the signal voltage.

Silicon controlled rectifier (SCR)   A three-electrode semiconductor device which conducts like a diode when a suitable signal is applied to a *gate* electrode. Used as a unidirectional current switch.

Software   Term used to describe the program used for a computer system.

Telemetry   Transmission of measurement data to a remote receiver, usually by radio or short-range inductive link.

Teletext   A system of information dissemination via television channels, using suitably adapted TV sets. The viewer can select 'pages' of information.

Terminal   A remote control console, usually a visual display unit (VDU) from which programs can be entered and run.

Thermistor   A semiconductor element with an electrical resistance which is very sensitive to temperature. In contrast to metals its resistance *decreases* with increased temperature.

Thyristor   See 'Silicon controlled rectifier'.

Transducer (electrical)   A device which produces an electrical output in response to a specific physical quantity, property or condition which is measured.

Transistor   A three-electrode semiconductor device capable of amplifying an electrical signal.

Triac   The equivalent of two thyristors connected in inverse parallel, used as a bidirectional current switch.

Videotex   A system of information retrieval from (and provision to) a databank, via a telephone or cable network, using a VDU terminal.

Visual display unit (VDU)   Terminal where the display is presented on a television type screen.

Volatile memory   Memory, such as a conventional RAM device, where the memory contents will be lost if the power is removed.

Voltage to frequency (V/F) converter   A device for converting an analogue signal into a train of pulses, the pulse rate being proportional to the magnitude of the analogue signal.

Word   A unit of data in a computer system which consists of a number of bits treated as a single entity. Normal word lengths are 4, 8, 12, 16 and 32 bits.

Index